IT DOESN'T TAKE A ROCKET SCIENTIST

IT DOESN'T TAKE A ROCKET SCIENTIST

Great Amateurs of Science

John Malone

John Wiley & Sons, Inc.

Published by John Wiley & Sons, Inc., Hoboken, New Jersey
Published simultaneously in Canada

For general information about our other products and services, please contact our
Customer Care Department within the United States at (800) 762-2974, outside the
United States at (317) 572-3993 or fax (317) 572-4002.

Wiley also publishes its books in a variety of electronic formats. Some content that
appears in print may not be available in electronic books.

ISBN 0-471-41431-X

Printed in the United States of America

10 9 8 7 6 5 4 3 2 1

This book is dedicated to the memory of
Paul Baldwin.

CONTENTS

INTRODUCTION

Amateur is a word whose meaning can shift according to context. In sports, for example, it is used to designate a person who competes just for the love of it, without getting paid. Olympic figure skaters originally were not allowed to earn any money from competing or performing in non-Olympic years, but that distinction was almost completely eroded during the last two decades of the twentieth century. Actors, on the other hand, are divided into amateur and professional categories on the basis of whether or not they belong to a union, a requirement for working professionally, but even an amateur actor can earn a living by performing only at summer theaters and dinner theaters that do not have union contracts. In the world of science, however, people are regarded as amateurs because they have not been professionally trained in a given discipline at an academic institution. If you do not have a degree, and usually an advanced degree, establishment scientists will regard you as an amateur.

In the modern world, you can still be regarded as an amateur in one scientific discipline even though you've won a Nobel Prize in another. A case in point is Luis Alvarez, who won the 1968 Nobel Prize in Physics for his work on elementary particles. Ten years later, he joined forces with his son Walter, a geologist, to postulate the theory that the extinction of the dinosaurs had been caused by a massive asteroid colliding with Earth. Many scientists initially scoffed at this theory, and the fact that Luis Alvarez was working in a field for which he had not been formally trained increased the level of skepticism. The theory was eventually accepted, on the basis of scientific evidence concerning high iridium levels in the geological strata at the presumed time of impact, 65 million years ago, as well as the discovery of an immense undersea crater off the Yucatan Peninsula.

Despite the annoyance Alvarez created among geologists and astronomers by meddling in their disciplines, however, he cannot be considered an amateur scientist in the sense that the main subjects of this book are. You will meet him here, briefly, in a surprising context, but the main subject of that chapter is Arthur C. Clarke, now one of the most famous of all science fiction writers, but then a young man without a college degree serving in the radar division of the Royal Air Force. Clarke wrote a short technical paper in 1945 that drew on several fields in which he had educated himself. Ignored at the time, the ideas set forth in that paper would eventually lead to a communications revolution. Clarke did go on to get a college degree after the war, but when he wrote this paper he was unquestionably an amateur. That was one reason why it was dismissed as "science fiction" by the few professionals who read it at the time.

Such dismissal is a common theme for the remarkable men and women whose stories are told in this book. Their ideas were ahead of their time, and they came from individuals who had no "credentials." In some cases, it is difficult to fault the professionals for failing to see the importance of such work. Who, in the 1860s, would have expected that an obscure monk, who could not pass the tests necessary for certification as a high school teacher, would be able to lay the foundations of a scientific discipline that would become one of the most important of the next century? But that is what Gregor Mendel achieved, planting generations of peas in a monastery garden, and analyzing the results in a way that would provide the basis for the science of genetics.

Mendel, of course, is now studied in high school and college biology courses. So is Michael Faraday, whose work on electromagnetism and electrolysis was crucial to a wide range of later scientific breakthroughs. Yet the drama of Faraday's extraordinary rise from uneducated London paperboy to the pinnacle of nineteenth-century British science is far less known than it should be. Others you will meet in this book are still little known to the general public. Henrietta Swan Leavitt, for example, one of several women known as "computers" who sorted astronomical plates at the Harvard College Observatory at the turn of the twentieth century, made a discovery about Cepheid stars that led to Edwin Hubble's proof that there were untold numbers of galaxies beyond our own Milky Way. Sitting at a desk in a crowded room, she provided a fundamental clue to the vastness of the universe. Grote Reber explored the universe from his own backyard in Wheaton, Illinois, where he built the first radio telescope in the 1930s. A self-taught French-Canadian bacteriologist, Felix d'Herelle, discovered and named bacteriophages in 1917 and set in motion the lines of inquiry that would lead directly to the revelation of the structure of DNA.

Some readers may be surprised to see the name of Thomas Jefferson here. Yet quite aside from his enormous political influence and architectural accomplishments, Jefferson was very much an amateur scientist. In his spare time he managed to carry out the first scientific archaeological excavation, using methods that are now standard in the field. Like Joseph Priestley, the dissident British clergyman who discovered oxygen, Jefferson's curiosity about the world led him in unexpected directions. Indeed, all the amateur scientists in this book share a great intellectual curiosity. What are those fumes from the brewery next door? What are those clear spots that keep showing up in cultures of bacteria? How could television signals be sent around the world? Does the static emanating from space mean anything? Curiosity is the hallmark of the amateur scientist. Professional scientists are curious, too, of course, but they have been trained to channel their curiosity in particular ways. In the history of science, the curiosity of amateurs has often been more diffuse, even wayward, but it sometimes produces results that leave the professionals in awe, and at other times provides the basis for an altogether new scientific discipline.

The very word *scientist* is relatively new. Until the end of the eighteenth century, people who investigated what things were made of and how things worked were called *natural philosophers*. Prior to the nineteenth century, most scientists were in a sense amateurs, although some were far more educated than others. During the twentieth century, the scientific disciplines became so sophisticated that most of them left little room for amateurism. No one can expect to develop new theories involving quantum physics without a great deal of professional training. Yet there are still a few areas in which amateur scientists can make a name for themselves. You will find the stories of two such individuals here: David Levy, the famous comet hunter, and Susan Hendrickson, a woman of immense curiosity who learned an entirely new discipline and discovered the most complete *Tyrannosaurus rex* fossil ever found. David Levy graduated from college, but never took an astronomy course. Susan Hendrickson never even attended college. Amateurs can still accomplish extraordinary things, even in the specialized technological world we now inhabit.

CHAPTER 1

Gregor Johann Mendel

The Father of Genetics

It is 1854. In the low hills just outside the Moravian capital, Brüun, there is a monastery with whitewashed brick walls surrounding gardens, courtyards, and buildings that are chilly even in summer. The fortresslike walls were built to protect its original inhabitants, Cistercian nuns, who took up residence in 1322. The nuns departed late in the eighteenth century, and the monastery lay empty for a while, falling into disrepair. It was taken over by a community of Augustinian monks in 1793—they had been displaced from the ornate building they occupied in the center of Brüun because Emperor Franz Josef of the Austro-Hungarian Empire wanted their jewel of a building for his own residence and offices.

By 1854, the monastery of St. Thomas had been headed by Abbot Cyrill Napp for several years. Within the Catholic Church, the Augustinian order had a reputation for liberalism, and Abbot Napp was particularly forward-looking. Born into a wealthy local family, he had very good connections with the leaders of secular society in Moravia, which were useful when the more conservative local bishop objected to the extent of the research taking place at the monastery. Since 1827, Napp had even been president of the prestigious Royal and Imperial Moravian Society for

the Improvement of Agriculture, Natural Science and Knowledge of the Country (popularly, the Agriculture Society), which
had been founded in 1807, the same year that Emperor Franz I
had decreed that the monks of St. Thomas and other local
monasteries would teach both religion and mathematics at the
city's own Philosophical Institute. Among the monks at St.
Thomas there was one for whom Abbott Napp had a particular
fondness, even though—or perhaps because—he was something
of a problem. That monk was Gregor Johann Mendel, and in
1854, with Napp's blessing, he began an experiment with garden
peas that would ultimately prove to be one of the greatest scientific breakthroughs in a century filled with them, providing
the basis for what we now call *the science of genetics*.

On the surface, there was little about Gregor Johann Mendel's life to suggest that he was remarkable. There were oddities
about it, but they appeared to indicate weaknesses rather than
strengths. Born in 1822, the middle child and only boy in a family that also included two girls, he grew up on a farm in Moravia, then under Austrian rule but now part of the Czech Republic. (Brüun is now known by its Czech name, Brno.) His father
Anton was extremely hard-working and tended toward dour-
ness, a trait even more pronounced in his older daughter, Veron-
ika. His wife, Rosine, and the younger daughter, Theresia, were
both of a contrasting sunny disposition. Gregor (who was chris-
tened Johann and assigned the name Gregor when he later
became a monk) alternated between his father's pessimism and
his mother's cheerfulness. Families everywhere, then as now, ex-
hibit character traits that appear to have been passed down from
parent to child, but in the twenty-first century we recognize that
some of those qualities of personality and mind-set are a matter
of genetic inheritance. Gregor Mendel himself would establish
the first scientific basis for that understanding, only to have his
work ignored in his lifetime and for fifteen years beyond it.

Gregor was a bright child, and ambitious. As a teenager, he
wrote a poem in celebration of the inventor of movable type,

Johann Gutenberg, which concluded with lines expressing hope that he, too, might attain the "earthly ecstasy" of seeing ". . . when I arise from the tomb / my art thriving peacefully / among those who are to come after me." There were impediments to any such grandiose achievement, however. The family's financial resources were modest, which would make it difficult to obtain a higher education. In addition, he was subject to periodic bouts of a psychosomatic illness that would keep him in bed for weeks at a time. His father and older sister had little patience with this kind of behavior, but his mother and younger sister indulged him. Theresia went so far as to give him her share of the meager family estate, which should have been her dowry, so that after graduating from the gymnasium (secondary school) he could go to the Philosophical Institute in the Czech-speaking town of Olomouc, a two-year program required of all students who wished to study at a university.

His sister's sacrifice would be repaid in later years, when he assisted her, financially and otherwise, in the raising of her three sons, two of whom would become physicians thanks to the help of their uncle. Yet even with his sister's loan, it was clear that there would not be enough money to attend university. Neither a modest scholarship grant nor his own efforts to earn money by tutoring would add up to sufficient resources. There was only one path open to him if he wanted a further education: he must become a monk.

Mendel was fortunate to have a physics professor at the Philosophical Institute, Friedrich Franz, who was himself a monk and an old friend of Abbot Napp at the monastery of St. Thomas. Even though Franz could muster only a modest recommendation concerning Mendel's intellectual ability, Napp agreed to take him in. Mendel arrived at St. Thomas in 1843, at the age of twenty, and spent the next five years studying to become a priest, starting as a novice and then moving up to subdeacon and deacon. He was moved through these steps more rapidly than would ordinarily have been the case, for the simple reason that the monastery had a shortage of priests. As Robin Marantz Henig explains in her book *The Monk in the Garden*, a number of monks who had administered last rights to patients at nearby St. Anne's Hospital had contracted fatal diseases themselves. Mendel was ordained as a priest two weeks after his twenty-fifth birthday, on August 6, 1847, and spent another year completing his studies before taking up pastoral duties. It quickly became apparent that he was far too shy and uncertain of himself to deal with parishioners. Indeed, he once again took to his bed, seriously ill without being sick. Abbot Napp decided that Mendel would be more usefully and successfully employed as a teacher, and the local bishop somewhat reluctantly sent him south to Znojmo to become an instructor in elementary mathematics and Greek at the secular gymnasium in that ancient town.

Mendel's year of teaching was a success. The discomfort he felt with adults he didn't know well, which had made pastoral duty so onerous, didn't affect him in dealing with youngsters, and he was also well regarded by his fellow teachers. He now had hopes of becoming a fully accredited high school science teacher. But in 1850, he failed the written and oral tests necessary for accreditation. Mendel's biographers have speculated at length about the reasons for this collapse. Part of the problem seems to have been a kind of "performance anxiety," no doubt connected to his tendency to psychosomatic illness. But there is also evidence that he sometimes refused to give the expected answers because he disagreed with current beliefs on a variety of subjects. In addition, there was a scheduling mix-up that meant the professors administering his oral exam had to meet on a date when they had expected to be free to travel, putting them in a foul mood. Six years later, however, when he tried again, his performance was even more dismal, and he seems to have simply given up after getting into an argument about an early question. What made this second failure profoundly discouraging was that he had spent two of the intervening years studying at the university in Vienna.

Mendel would continue teaching at the grade school level for a number of years, but his failure to gain more substantial academic credits underscores some important points about the nature of scientific amateurism. As we will see throughout this book, great amateur scientists have often received a considerable amount of education, but it tends to be spotty and sometimes lacking, ironically, in the very area in which the scientist ultimately makes his or her mark. In Mendel's case, he received more mathematics training than anything else, and that would make it possible for him to apply a mathematical rigor to his experiments with pea plants that was highly unusual for the period. He also studied some botany, but this was a subject that caused him particular problems. Because he was a farm boy, he had an ingrained knowledge of plants that caused him to balk at various academic formulations. When a student refuses to give the answer he or she has been taught, academicians inevitably conclude that the student is stupid rather than reassess their own beliefs. Brilliant amateurs have always been prone to question the questioner, and that usually gets them into deep trouble.

The end result is often a young person of great talent who has not attained the kind of academic degree or standing that would serve as protection when he or she puts forward an unorthodox idea. Even the attainment of academic excellence may not be enough to stave off attacks from establishment scientists; without such achievements, new concepts are likely to be utterly dismissed. There is another side to this coin, however. If Gregor Mendel had in fact passed the tests that would have made him a full-time high school teacher, it is unlikely that he would have had the time to devote to the experiments that would eventually make his name immortal.

———

Between attempts to gain accreditation, Mendel also began his first experiment in heredity, which predated his efforts with peas.

He was allowed to keep cages of mice in his quarters, and bred wild mice with captive albinos in order to see what color successive generations would turn out to be. Selective breeding of both animals and plants had been practiced for centuries by farmers like his father, but even though a farmer might succeed in improving the strength of his animals and the hardiness of his plants, no one had any idea why or how such improvements occurred. Mendel wanted to know exactly that. Although the Catholic Church now recognized the importance of scientific inquiry in general, not all its leaders were happy about this trend. It was true that the Church had embarrassed itself in forcing Galileo to recant his belief in the Copernican model of the solar system in 1638 (an apology was finally issued by Pope John Paul II in 1998), and had gradually found ways to reconcile scientific progress with its theology, but there were some who found it unseemly for members of the priesthood to be involved in such matters. One of those who took a dim view of scientific research was the local bishop, Anton Ernst Schaffgotsch. He was more or less at war with Abbot Napp for many years, but the abbot had too many prominent local friends, and his monks were too highly regarded as teachers, for the bishop to get away with closing down the monastery, as he would have liked. Nevertheless, he was able to set some limits, and during one confrontation with Napp he decreed that Mendel's mice had to go. He was particularly disturbed that sexual congress was at the heart of the monk's experiments.

Without knowing it, the bishop actually did Mendel a great favor. While mice were regarded as very simple creatures with obnoxious habits, they are in fact genetically complex. We now know them to be biologically similar to humans in many ways, which is one reason why they are so often used in medical experiments. If Mendel had continued to experiment solely with mice, it would have been impossible for him to achieve the breakthrough he did. The very complexity of the creatures would have derailed his project.

And so, in 1854, Mendel turned to the common pea. There had been an experimental garden at the monastery for more than two decades, and such work was seen as a potential benefit to agriculture in general, and far more seemly than breeding generations of animals. Mendel is reputed to have commented, with amusement, that the bishop failed to grasp that plants also had sex lives. The reproductive mechanisms of plants are in fact quite varied. Some species have specifically male and specifically female plants. If the gardener does not make sure to have a male and a female holly bush, for example, and to plant them near one another, there will be no berries. A great many plants depend upon bees for pollination—if the bee population is destroyed in a locale, numerous plants will die out, having no way to reproduce. The common garden pea, the species *Pisum,* that Mendel experimented with would survive the loss of the bee population, however, since they are hermaphroditic, each flower containing both the male *stamen* and the female *pistil.*

The fact that peas are hermaphroditic was important to Mendel's experiments, because it made it possible for him to exercise complete control over their reproduction. Such control did require a great deal of painstaking work. The yellow pollen that contains the male gamete (sperm) is produced in the tiny bulbous anther at the top of each antenna-like stamen. Under usual circumstances, the pollen will fall onto the sticky stigma of the female pistil, and pass down the canal known as the style to the ovules (eggs). In order to crossbreed different pea plants, the monk had to proceed slowly down a row and remove the pollen by hand from the stamens of plants he wanted to fertilize with the pollen of another. He was in effect castrating each plant on which he carried out this operation. He would then cover the buds with tiny caps of calico cloth, to protect them for the few days it would take for the female stigma to mature and become sticky. The cap also prevented any insects from fertilizing a castrated plant with the pollen from still another plant. When the stigma was mature, Mendel would pollinate it with the gametes gathered from another plant with different characteristics.

We do not know how Mendel kept track of what he was doing. No logbooks or notes exist, only the final paper he would present to the Agriculture Society in two sections, a month apart, in 1865, which was then published by the Society. All his other papers were burned in the courtyard of the monastery following his death—but that is getting ahead of the story.

What we do know from the 1865 paper reveals an extremely orderly mind, and an entirely new way of categorizing the results of crossbreeding experiments. There is a language problem that needs to be cleared up before we look at the experiments in more detail, however. In Mendel's day, crossing any organism with another was called hybridization. No distinction was made between crossing organisms of two different species and crossing organisms that were merely different varieties of the same species. Mendel's two-part paper of 1865 was titled "Elements in Plant Hybridization," but today that title would be regarded as incorrect, since he was in most cases crossing varieties of the same species of peas. Today, the creation of a true hybrid is defined as the crossing of different species, as the tangerine and the grapefruit were crossed to create the tangelo. Among animals, a mule is a hybrid of a horse and a donkey, and the mule is sterile, as is often the case with hybrids, although in plant hybridization fertility can be restored by chemical treatment that doubles the chromosomes.

Mendel's work did not suffer from the confusion surrounding the meaning of hybridization, however. He determined that garden peas had seven distinct characteristics, or traits, that were always exhibited in one of two ways, as can be seen in the following chart.

The Seven Traits of Pea Plants

TRAIT	VARIETY
Seed shape	Smooth or wrinkled (alternatively round or angular)
Seed color	Yellow or green
Seed coat color	White or gray
Stem length	Tall or short
Shape when ripe	Pods inflated or constricted
Color of unripe pods	Green or yellow
Position of flowers	All along stem or single at top of stem

There are a few aspects of this list that require special comment. Many books use only the descriptions "smooth" and "wrinkled" in respect to seed shape. But as Henig makes clear in *The Monk in the Garden*, neither smoothness nor being wrinkled are really a matter of shape. In her view, and that of other specialists, a mistranslation from the German is at fault, and what Mendel was really looking at were actual shapes, round and angular. In addition, the third characteristic, seed coat color, sometimes appears as flower color. He did start out with flower color, but apparently realized that flower color was linked to other characteristics, and thus added, and paid special attention to, the color of the seed coat. We now know that each of the seven traits listed above, including the white or gray color of the thin translucent seed coat, is determined by a separate chromosome and transmitted independently.

Before starting to cross different varieties of pea plants, Mendel spent two years growing plants from the seeds produced by each variety. This was done to make certain that all the plants he was using were "true," and would not produce any variations on their own. The fact that he spent so much time laying a rigorous foundation for his experiments is one of the reasons his work has come to be so highly regarded. Many people might think this was a boring prelude to the experiments to come, but Mendel took so much pleasure in gardening for its own sake that even this preliminary stage must have brought its satisfactions.

Once he was certain that the plants were stable down through several generations, he began crossing plants carrying each of the seven traits with other plants carrying the opposite trait. Plants that produced round seeds were crossed with plants bearing angular seeds, tall plants with short ones, single-flower stems with multiple-flower stems. At the time it was believed that heredity was always a matter of a balance being struck— thus the crossing of a tall plant with a short one would be expected to produce a medium height plant. But that was not what happened. The crossing of a tall plant with a short one always produced tall plants in the next generation. Nor did the crossing of plants with yellow pods and plants with green pods produce a new generation that was a greenish-yellow blend— instead the pods were all green. From these results, which held true for all seven traits, Mendel came to the conclusion that some "factors," as he called them, were stronger than others. The stronger factor he called *dominant*, the weaker factor he called *recessive*. Those two terms are still in use today, a tribute to their aptness and to Mendel's genius. He did not know what the factors themselves were, however—it would not be until 1909 that the Danish professor of plant physiology Wilhelm Johannsen coined the word *gene* to describe these factors, and it would take until the 1940s to determine that genes could be identified with a particular length of DNA, the complex molecule that contains the chemically coded information necessary to make proteins.

Year after year during the remainder of the 1850s and on into the next decade, Mendel crossed his pea plants. Some he certainly grew outdoors in warm weather, but others must have been raised in colder months in the two-room glasshouse in the monastery courtyard next to the brewery, and later in the greenhouse that was built on the orders of Abbot Napp for Mendel's particular use. The small glasshouse was heated by a stove, but the larger greenhouse, erected in a sunnier location, was warmed only by the heat of the sun. Robin Marantz Henig relates in great detail the arguments that developed in the twentieth century about the exact location of the outdoor garden where Mendel grew his peas. Those arguments occurred because the officially designated plot seemed too small and too shady to have sustained all the plants Mendel said he had grown. This constricted, sun-deprived plot of land seemed to some doubters evidence that Mendel had lied about the extent of his research. Only in the 1990s years did extensive scholarly detective work establish a larger, sunnier plot of land by the greenhouse as the location of his main garden. The clue that solved the mystery revolved around which windows his fellow monks would call out to him from as he worked with his pea plants. It was long assumed that they had greeted him from the windows of the formal library that overlooks the smaller plot, but in fact the monks spent most of their time in the study rooms at the far end of the structure, where the windows opened onto an entirely different part of the grounds.

After Mendel had tested many generations of pea plants following the initial crossing for each of the seven traits, he moved on to cross these plants again. He expected this double crossing to once again affirm the strength of the dominant factors, including tallness and green pods. To his astonishment, that was not the result. Some of the plants turned out as expected, but others did not. A lesser man might have thrown up his hands in despair at this point. Was his theory about dominant and recessive genes incorrect, after all? But the monk had been putting his mathematical training to use from the start, and now those carefully kept figures revealed an even greater secret. Again and again, the new plants produced a 3:1 ratio—for every three plants that did retain the dominant gene, one did not. This ratio held true for all seven traits. Such a ratio could not be mere accident. Something profoundly ordered was at work, and Mendel's 3:1 ratio would become the basis for the work of hundreds of twentieth-century scientists seeking to unravel the full story of the genetic code of living organisms, including human beings.

———

Mendel was not working in a complete vacuum as he carried out his experiments on pea plants. In the 1730s, the Swedish botanist Carl von Linné, writing under the Latinized name Carolus Linnaeus—by which he is primarily known today—created a system for categorizing all living things, divided into two "kingdoms," plant life and animal life. Within each kingdom, organisms were subdivided, in descending order from the broadest to the most specific groups, into classes, orders, genera, species, and finally the varieties within a species. This system has become more complicated with the development of further knowledge, so that we now have five kingdoms instead of two, and a phylum (for animals) or a division (for plants) that precedes the class. There are also possible subclasses, which precede the order, and families, which precede (and sometimes coincide with) the species. Human beings belong to the chordate kingdom, the mammalian phylum, the primate class, the hominid order, and the family/species homo sapiens. Even with the less complex system devised by Linnaeus, however, the seeming chaos of nature was given shape in a way that made it possible for anyone, whether knowledgeable amateur or professional scientist, to understand exactly what plant or animal was being described.

Yet Mendel did not in fact know the exact classifications of his peas. They were all common garden peas, of the genus *Pisum*, some already growing in the monastery garden and some that he sent away for. He believed that most were *Pisum sativum*, although experts suspect that some other species aside from *sativum* were among his specimens, such as *Pisum quadratum*. Mendel was a bit cavalier about this question, feeling that in terms of what he was interested in doing, it didn't much matter provided he made certain at the beginning that each plant bred true. His lack of concern about the exact species of each plant was not surprising—he had gotten into trouble on his exams about precisely this kind of detail. But he was correct—it did not really matter in respect to his particular experiments. A more "professional" or academic scientist might well have gotten hung up on the fact that the exact species of each plant was not known, but Mendel's "amateurism" in regard to this matter allowed him to proceed enthusiastically with the more crucial two-year testing period to make sure each plant bred true generation after generation. Amateurs can get things terribly wrong by ignoring "academic" details, but the brilliant amateur can sometimes vault over a problem by virtue of his or her recognition that the "correct" way of proceeding may not be necessary in a given situation.

When Mendel began his experiments, there was also a great deal of ferment about the subject of transmutation, which would soon come to be called *evolution*. This hubbub had begun in 1809, with the publication of a book by the French naturalist Jean Baptiste de Lamarck titled *Philosophie zoologique*. Lamarck coined the word *biology* and was the first to distinguish vertebrate from invertebrate animals (leading to a major addition to the categories devised by Linnaeus). But his reputation suffers from the fact that his ideas about evolution turned out to very wrong, and were ultimately seen as ridiculous. He believed that plants and animals changed according to their environment, which was an accurate enough supposition, but his examples of how they changed now sound like the fables in Rudyard Kipling's *Just So Stories*, such as "How the Elephant Got His Trunk." No matter how flawed, though, his work proved a bombshell that appeared to call into question God's place in creating the creatures of the earth. The idea that one hungry giraffe stretched his neck to eat leaves that were seemingly out of reach higher up a tree, and that the results of such stretching would be instantly fixed, and thus passed on to the giraffe's offspring, seemed blasphemous rather than silly to the devout, including many scientists. Thus there was consternation regarding evolutionary concepts even before the publication of Charles Darwin's *Origin of Species* in 1859, when Mendel's own experiments were still four years short of completion. Darwin rejected Lamarck's ideas, being influenced instead by the economist Thomas Malthus's concept of a "struggle for existence," which he transformed into the "survival of the fittest." But because Darwin believed, in contrast to Lamarck, that change took place in plants and animals over very long periods of time, his work also flew in the face of biblical dogma concerning the creation. More blasphemy, according to many.

Mendel clearly became familiar with Darwin's work, since he would send him a copy of his two-part lecture on his pea crossings in 1865. But he managed to present his own work in a way that avoided direct entanglement in the great evolutionary debate. The implications of Mendel's experiments certainly had importance in terms of evolution, and he must have realized that they did, but his mathematical approach was so new and so dry that it obscured the controversy lying below the surface of his numbers. The truth is that virtually no one understood what he was doing. To the extent that his work seemed in any way remarkable to those who heard his lectures or read the published version, it was largely a matter of simple amazement that he could keep such close track of all those thousands of pea plants grown, generation after generation, over so long a period of time. "So much work—he must be quite clever," appears to have been the general reaction.

In the latter stages of his experiments, the work became even more complicated. He had established that the double and triple crossing of his plants would produce a 3:1 ratio between dominant and recessive factors. But to establish conclusively the nature of the dominant and recessive factors, it was necessary to take a further step, backcrossing his hybrids with original parent plants in two different ways. Half these crosses were made with double dominant plants, half with double recessive plants. He expected that the double dominant crosses would produce plants that were all alike in appearance—the dominant factor would be so strong as to mask any underlying recessive factor. On the other hand, the results from the double recessive crossings ought to be four different types in a 1:1:1:1 ratio, because the recessive factors would not be suppressed by any dominant ones, and would therefore resurface. That was exactly what happened.

It would be another half-century before the technical language would be developed to explain these results. But Mendel had clearly demonstrated the difference between a *phenotype* (in which the physical traits are visibly displayed) and a *genotype* (in which the gene variants are present, and still capable of being passed on to another generation, but are not necessarily visible). He had started with a theory and ended with confirmation of what eventually came to be called Mendel's laws. He went on to experiment for a couple of years with a variety of other plants, including snapdragons and maize, which appeared to show that the results he had achieved with his peas would hold true for any plant.

The pea experiments were concluded in the summer of 1863. That turned out to be extremely fortunate, since the next year almost all his pea plants were destroyed by the pea weevil. If that pest had shown up three or four years earlier, it would have been impossible to carry out his backcrosses of the double dominant and double recessive plants. Mendel's physical condition by 1863 was also making his work more difficult. His eyesight was getting poorer, and he was quite heavy—the latter the result of a monastery kitchen widely known for the excellence of its cook, Luise Ondrakova, who would eventually write a cookbook containing, among many soups, strudels, and pork dishes, her famous rose-hip sauce for meat.

For two years, Gregor Mendel worked on the paper describing his experiments. He delivered it in two parts, on Wednesday, February 8, and Wednesday, March 8, to the Brüun Agricultural Society. There are conflicting reports about its reception, but it was at the least polite. The society duly published the forty-four-page paper, and Mendel ordered forty copies of it, which he proceeded to send out to many of the most illustrious scientific names in Europe, including Charles Darwin. A number of these copies were found in later years, when Mendel's work was rediscovered. But at the time almost no one paid real attention. In those days, such publications arrived with the pages folded over. In order to read them, it was necessary to cut the pages. Darwin's copy, and some others, were not even cut. They had never been read by the recipients.

Mendel's dismay that there was no reaction from thirty-nine of the important scientists to whom he sent his paper was offset by what he considered the importance of the one reply he did get. When he was studying in Vienna, one of his teachers, Franz Unger, often praised the work of Karl von Nägeli, a professor of botany at the University of Munich. In 1842, Nägeli had described the processes of what we now call cell division and seed formation in flowering plants. Mendel became almost obsessed with Nägeli, and sent him a copy of his paper, together with an explanatory note, at the very end of 1866. It was two months before he received a skeptical reply—several drafts of which, we now know, had been composed by Nägeli. The professor held to the belief that crossbreeding produced a blend, and he appears to have recognized that if Mendel's experiments were correct, it would prove that view wrong. Thus he suggested that the experiments had not been carried far enough, and that even though they might be correct as far as they went, they did not provide sufficient justification for any general law. Mendel wrote back, trying to further clarify certain aspects of his paper. There was no answer to that letter, and Mendel tried a different approach in a third and then a fourth letter, finally suggesting that he could act as a kind of assistant to Nägeli in his own experiments, if the professor would send him some seeds to work with. That got a response, and the correspondence continued intermittently for seven years. Unfortunately, the seeds that were sent to Mendel were hawkweed (*Hiercium*), and crossbreeding them was fruitless, because hawkweed usually reproduced in a way that produced clones, called *apomixis* in plants and *parthogenesis* in animals. There is some question to this day about whether Nägeli knew he was giving Mendel an insoluble problem. The frustrating results, at any rate, even led the monk to question his own previous success with peas.

But Mendel would have less and less time for experiments anyway. On March 30, 1868, he was elected the new abbot of St. Thomas on a second ballot, succeeding Abbot Napp, who had just died at the age of seventy-five. He now had a great many administrative and social duties to occupy his time. As though to put a final exclamation point to his years of experiments, a freak tornado in October 1870 destroyed the greenhouse that had been built for him. He continued to serve as abbot until his death on January 6, 1884, but his standing in the community waned due to an endless tax dispute with the government.

Anselm Rambousek, whom Mendel had defeated in 1868, then became the new abbot, and soon saw to it that his predecessor's papers were burned. It would be another fifteen years before Mendel's work was rediscovered. His name was not unknown—he was listed fifteen times in Wilhelm Obers Focke's work on plant hybridizing, and because of those references, he was accorded a brief mention in the next edition of the *Encyclopedia Brittanica*. But neither publication made clear the significance of his work. That would have to wait until 1900, when three biologists almost simultaneously came across Mendel's original paper.

A botanist from Amsterdam, Hugo de Vries, was working along lines similar to Mendel's, but with different plants, when he came across the monk's paper, probably in 1899. Mendel's work backed up his own, but of course it also anticipated it by a quarter-century. De Vries made use of Mendel's terminology in a lecture but did not credit Mendel. The lecture was published and read on April 21, 1900, by his rival Karl Correns. Correns was also working on the question of hybrid relationships, and was infuriated by the fact that de Vries had once again beaten him to the punch, and not properly credited Mendel in the bargain. Correns happened to be married to a niece of Nägeli, and was able to gain access to the correspondence between the two men. The third rediscoverer, ironically, was the grandson of one of the professors, Eduard Frenzl, with whom Mendel had argued during his second failed attempt to gain accreditation as a high school science teacher. Young Eric von Tschermak published a paper in June 1900, in which he tried to anoint himself the true "rediscoverer" of Mendel, although many experts feel he never fully understood Mendel's work.

Correns wrote an essay of his own, titled "G. Mendel's Law Concerning the Behavior of the Progeny of Varietal Hybrids," in which he came close to accusing de Vries of plagiarism. Perhaps hearing about Correns's essay in advance, de Vries belatedly mentioned Mendel in a footnote added to a German translation of his lecture, but also tried to suggest that he had arrived at his own conclusions before coming across Mendel's thirty-five-year-old paper. The motivations of all three of these men have been debated ever since. Jealousy and self-aggrandizement certainly played their part, but as a result the name of Gregor Mendel was suddenly a very hot topic indeed.

In the end it was not any of these three men who would serve as the chief promoter of Mendel, however. That role was taken by a zoologist at St. John's College, Cambridge, who had turned thirty-nine in 1900, William Bateson. The historian Robert Olby has suggested that Bateson may have later fictionalized his own recognition of Mendel's genius, in order to make it more dramatic. He knew de Vries well, and would obviously have read his lecture—in the German translation that included the footnote mentioning Mendel's name—and there was a copy of Mendel's original 1866 paper in the Cambridge University library. But there are questions as to whether he could have gotten hold of Mendel's paper in time to read it, as he claimed, on a train ride to London on May 8, 1900. Bateson was supposed to give a lecture that day, and claimed to have revised it on the spot in light of reading Mendel's paper, but accounts of the actual lecture do not even mention Mendel. Nevertheless, Bateson did become Mendel's main champion.

Over the next several years, Mendel became the center of a bitter argument between two schools of thought. One school, following Darwin's lead, held that evolution occurred slowly, and in a continuous curve. Bateson believed that it occurred in discontinuous leaps, and that Mendel's laws showed how that could happen. The fight between these two groups continued for a decade, sometimes in the form of published papers, sometimes in public debates. In the midst of this scientific turmoil, Bateson invented the word *genetics,* although oddly enough, the word *gene,* to describe Mendel's "factors," did not come into use until several years later. The details of the debate between the Mendelians and the followers of Darwin, who were known as *biometricians,* are highly technical, but ultimately one major scientist after another came over to the side of the Mendelians, for many particular reasons and one general one: Gregor Mendel's laws proved to be the most useful and logical approach to the new science of genetics.

In his later years as the abbot of St. Thomas, when the subject of his pea experiments came up, Mendel sometimes said to friends, "My day will come." He said it gently, even humorously, by all accounts. He was not a man with a large ego. Those who would fight in his defense long after his death often had outsize egos, and reputations to protect, which meant that the debate could become extremely heated. The Moravian monk was an amateur who tended his rows of peas with unflagging devotion and care for nine years. Abbot Napp must have had some sense that the monk who could not pass his teaching exams or deal well with parishioners (at least in his younger years) had something very special to offer, or he would not have ordered the greenhouse to be built. But no knows whether Napp truly understood the importance of what Mendel was doing, or had any real inkling that the young monk was a genius.

Indeed, there have always been some scientists who have questioned whether Mendel deserves as much credit as he now gets. There have been claims that his results were too "perfect" and that he must have fudged the numbers. The long debate over the exact location of his garden—whether in the shade or in the full sun—was fueled by the annoyance of some scientists that a mere amateur should be credited with the foundation of a discipline that became one of the greatest success stories of the twentieth century and seems destined to be even more important in the twenty-first. DNA research, the ongoing effort to map the entire human genome, cloning, genetic modification of fetuses to banish inherited diseases and create more nearly perfect future humans—that all those headline-grabbing aspects of genetics should be traced back to a monk growing peas in a monastery garden is galling to some professionals. But any doubts have been overwhelmed by the fact that Mendel's laws hold true. A mere amateur began it all, while many of the great minds of his own time were on the wrong track. Rows of peas, in a sunny garden, in a special greenhouse, tended for nine years by an increasingly fat monk who all alone began a revolution in human knowledge.

To Investigate Further

Henig, Robin Marantz. *The Monk in the Garden*. New York: Houghton Mifflin, 2000. A National Book Critics Circle Award finalist, this is one of those rare books that manages to get the facts right and tell a scientific story with great charm at the same time.

Stern, Curt, and Eva R. Sherwood. *The Origin of Genetics: A Mendel Source Book*. San Francisco: W. H. Freeman, 1966. This volume contains translations of Mendel's paper, his correspondence with Nägeli, and other documents. It is regarded by many experts as the best translation of Mendel. Although out of print, it can be found in many libraries.

Orel, Vitezlav. *Gregor Mendel: The First Geneticist*. Oxford and London: Oxford University Press, 1996. This was the first major Mendel biography since 1926. Henig draws on it extensively, but her own book is probably the better bet for general readers.

Tagliaferro, Linda, and Mark V. Bloom. *The Complete Idiot's Guide to Decoding Your Genes*. New York: Alpha/Macmillan, 1999. For those who like to absorb complex subjects in small bites, this is an excellent overall introduction to genetics, with an early chapter on Mendel, and final ones that deal with modern controversies such as cloning.

CHAPTER 2

David H. Levy

Comet Hunter

On July 16, 1994, telescopes all over the Earth, as well as the orbiting Hubble Space Telescope and the cameras of the *Galileo* spacecraft, were trained on the planet Jupiter. The twenty-one pieces of Comet Shoemaker-Levy 9 would begin to crash into the atmosphere of Jupiter that day. No one was certain whether the impact would produce the most spectacular cosmic event since Galileo built the first true telescope in 1610 or whether the result would be a complete fizzle. But if there was to be a great show, no astronomer wanted to miss it.

The chances seemed good that there would be at least something to see. Even before it had been first photographed on March 23, 1993, with the Mount Palomar eighteen-inch (0.46 m) telescope-camera, the comet had broken into pieces of various sizes. That had occurred, it was later calculated, when the comet had a near miss with Jupiter on July 7, 1992, passing the giant planet with a mere 13,000 miles (20,000 km) to spare. Torn apart by Jupiter's immense gravity, the comet continued on its tightening orbit around the planet. Some of the pieces of the comet, it was believed, might be as large as three kilometers across. That was larger than the asteroid that formed the Chicxulub crater on the coast of the Yucatan Peninsula on Earth 65 million years ago—a collision that most scientists had come to accept as the chief reason for the extinction of the dinosaurs. On the other hand, Earth is a terrestrial planet, and when a large asteroid hits it, the amount of dust thrown into the atmosphere would be sufficient to obscure the sun for as much as two years. Jupiter, however, is a gas giant eleven times the size of Earth, and it was conceivable that it would be able to simply swallow the pieces of the comet, no matter how large, with scarcely a ripple on its vast surface.

As astronomers waited for the impact on July 16, 1994, perhaps no one was as nervous or as excited as David H. Levy, who shared the comet's discovery with Gene and Carolyn Shoemaker. It was the ninth comet they had discovered together, hence its designation as Shoemaker-Levy 9. While it was the most important comet by far that this team had been the first to spot, and thus a major feather in everyone's cap, it held special meaning for Levy. Although it was the twenty-first comet Levy had discovered on his own or in conjunction with others, this one was a celestial object on a scale and of an importance that even professionals usually can only dream about. For an amateur like David Levy, having his name attached to it was the culmination of a lifelong fascination with comets.

Levy was born in Montreal, Canada, in 1948. In interviews, he has recalled being introduced to astronomy in the second grade, when friends gave him a copy of *The Big Book of Stars*, which was only twenty-six pages long but seemed huge to him. His mother, Edith, a physician, and his father, Nathaniel, a businessman, both encouraged their son's early interest in astronomy. In the sixth grade, he gave a speech at school on Halley's Comet. Although he had decided he would become an astronomer by the time he was twelve, and was given his first small telescope by an uncle that year, that ambition was superseded by his interest in English literature. He attended Acadia University from 1968 to 1972, gaining a B.A. in English, followed by a master's degree at Queen's University in 1979. Although he did much less observing during these years, the subject of his master's thesis on Gerard Manley Hopkins was titled "The Starry Night: Hopkins and Astronomy." Levy is often asked by reporters and television interviewers how many courses in astronomy he has taken. His answer is wryly given an honored place on his web page (www.jarnac.org). A list of links includes the heading, "*Here is a list of all the astronomy courses David has taken.*" Clicking on the link takes you to a very brief statement: "I have *never* taken an astronomy course."

Levy is obviously proud of being an amateur. But it is not a matter of being proud because he has become a very famous amateur. He has always felt that amateur astronomers are a particularly congenial group, the kind of people who become knit together as members of an extended family. Beyond that, in an interview with Robert Reeves for the magazine *Astronomy*, he noted, "You do have some advantages as a backyard astronomer. As an amateur, you have more freedom. Nobody expects you to discover something. Nobody expects you to come up with exciting theories or career-sustaining developments."

Levy began looking for comets with a three-inch (7.6 cm) backyard telescope in 1965, on December 17. For years, although he observed comets discovered by others, he found no new ones of his own. In 1979, he moved to Tucson, Arizona, because the sky there was so much darker—in Montreal, as in the vicinity of any large city, electric lights wash out the details of the night sky. But he spent nearly five years in Tucson before he found his first new comet, on November 13, 1984, which was officially reported the next day. It had taken nineteen years and a total of 917 hours of observing to arrive at that first success. Clearly, Levy is to be believed when he says that the amateur astronomer observes the sky for the pure love of doing it.

That first sighting of a new comet, which was named for him and the Soviet energy expert, Yuri Rudenko, also an amateur astronomer, was discovered with a backyard telescope. He would not spot another one until January 5, 1987, but then his backyard endeavors began to pay off more regularly, and over the next seven years he shared two other discoveries, and made four solo ones as well. His backyard discoveries are:

- Comet Levy-Rudenko, 1984t, Nov 14, 1984
- Comet Levy, 1987a, January 5, 1987
- Comet Levy, 1987y, October 11, 1987
- Comet Levy, 1988e, March 19, 1988
- Comet Okazaki-Levy-Rudenko, 1989r, August 25, 1989
- Comet Levy, 1990c, May 20, 1990
- Periodic Comet Levy, June 14, 1991
- Comet Takamaizawa-Levy, April 15, 1994

Designations such as "c," "e," or "r" indicate that according to official international records, this was the order, from a–z, of all comets found by amateur or professional astronomers in a given year.

The most important of these backyard discoveries was Comet Levy, 1990c. David Levy was fortunate to be the first to see it, since it was no ordinary comet. During the summer of 1990, it put on an extraordinary show, and was more easily seen from greater areas of Earth's surface than any comet since the 1986 reappearance of Halley's Comet. Halley's Comet would be back again, like clockwork, in 2062, but Levy 1990c would not be seen again from Earth for thousands of years, if ever. Where had it come from, and where was it going?

————

Most scientists believe that comets were created at the same time as the rest of the solar system, 4.6 billion years ago. During the early history of Earth, asteroids and comets existed in great numbers, and crashed into our planet almost nonstop, sometimes being absorbed by the still molten sphere of Earth, sometimes tearing off pieces of the new planet, pieces that would then become the basis of still more asteroids, and in some cases, comets. Asteroids and comets, however, are very different from each other. Asteroids are rocky chunks, often strangely shaped but still solid. The nature of comets remained elusive until the middle of the twentieth century. In 1950, the American astronomer Fred L. Whipple first proposed the idea that comets were essentially dirty snowballs, composed of a mixture of frozen water, methane, and carbon dioxide within which an enormous conglomeration of pebblelike material is held in suspension. Whipple's concept has held up over the decades, with some changes. It is now thought that comets contain less solid material than originally proposed, and it has been established that some comets have a large amount of methane as part of the mixture while others are low in methane. Like the combinations of snow and gravel shaped by the hands of young human bullies in winter, comets can be lethal if they strike another body, although they themselves are demolished in the process.

During the first several hundred million years of the existence of our solar system, comets and asteroids must have collided with the planets constantly, gradually ridding the planetary region of most of them. As many as 100,000 small asteroids still exist in a belt between Mars and Jupiter, but most comets appear to originate in the Oort Cloud, named for the Dutch astronomer who postulated it in 1950. The Oort Cloud lies beyond Pluto, 1.5 light-years from the sun, and it may contain as many as ten trillion comets. Occasionally, disruptions caused by the passing of other stars sends one of these comets swinging in toward the sun. The great majority of comets have such elongated orbits that once they have made their journey around the sun (at which point they may become visible from Earth), they travel so far out that they will not be seen again for thousands, perhaps millions of years. That is one reason why the great majority of the more than 800 comets whose orbits have been calculated will not be seen again from Earth for millennia to come. However, about 160 of these comets are periodic in nature. They have been trapped by the gravity of one of the larger planets, usually Jupiter, or the sun itself, and will reappear in a time frame ranging from as much as 200 years to as little, in one case, as 3 years.

When a comet approaches the sun, the ices that hold it together start to vaporize, and the gases that are released inflate the coma, or nucleus, from a modest few kilometers in diameter to a size that may equal that of Earth itself—and that, of course, makes them potentially visible to us. A dust tail forms, sometimes fan-shaped, sometimes forming long, curving streaks. An ion, or gas, tail may also develop, and these tails can change their appearance rapidly, just as the coma itself may send off spectacular jets of erupting material.

Because very large comets were often visible to the naked eye, they were regarded as portents of good or evil throughout most of recorded history. They might be interpreted as warnings that a crop failure or the fall of a royal dynasty was imminent. Occasionally, they might come to be associated with the dawning of a new epoch and a new hope—many scholars believe that the "star" that guided the Three Kings to the stable where Jesus was born in Bethlehem was in fact a comet. No one, of course, had any idea what these bright, fast-moving objects in the night sky really were during most of human history. There are about 400 comets of which we have recorded evidence before 1610 and the invention of the telescope. It would be the theories of Isaac Newton, however, that would provide his friend Edmund Halley with the mathematics necessary to gain a much greater understanding of the paths of comets through the solar system.

In a period of a mere two years, from 1665 to 1667, Isaac Newton invented differential and integral calculus, proved that white light actually consisted of a rainbow of colors, and discovered the three standard laws of motion as well as the universal law of gravity. The publication of his findings on calculus was met with a claim by the German mathematician Gottfried Leibniz that *he* had discovered calculus—in fact, the two men were working on the same problems simultaneously, unknown to each other, and Leibniz, without being aware of what Newton was doing, finished his own work just after Newton and before the latter's papers were published. But Newton was convinced that Leibniz had stolen his ideas, and resisted publication of any of his other discoveries for twenty-one years. Edmund Halley, who would eventually become the Astronomer Royal, spent years trying to persuade Newton to put his *Principia Mathematica* into publishable shape, and finally ended up paying for the printing of his friend's great work himself. But Halley was repaid in a way that ensured he, too, would be known down the ages. Halley believed that the great comets whose appearances had been recorded in 1456, 1531, and 1607, and the one he himself had seen in 1682, were in fact a single comet making a predictable return. Using Newton's work, he was able to calculate a parabolic path through the solar system that would bring it back once again in 1758, sixteen years after his own death. His prediction proved correct, and his name has been attached to the comet ever since.

Halley's work brought comet watching in general a new scientific respectability, although superstition dies hard. Even the 1910 appearance of Halley's Comet was surrounded by a hysterical uproar when a French scientist named Camille Flammarion suggested that the tail of the comet would dip into Earth's atmosphere, releasing a poisonous gas called *cyanogen,* which might contaminate the atmosphere sufficiently to kill all living things. Panic led some people to stuff towels into the cracks beneath their doors, a mania that for some reason took particular hold in Chicago. Nothing happened, of course, although there was a significant death associated with the 1910 visit from the great comet. Mark Twain, who had been born in 1835, when the comet last appeared, was not well, and predicted that having come in with Halley's Comet, he would also go out with it. He died on April 21, 1910, the day following the comet's brightest appearance in the sky.

The lore of comets down through the ages is such that many people become enthralled with them, whether as a child, like David Levy, or in adulthood, like Gene and Carolyn Shoemaker. During the 1980s, as Levy's name started being attached to comets he had discovered with his six-inch backyard telescope, he ran into the Shoemakers at various conferences, and hit it off extremely well with them. They had arranged to use the smaller eighteen-inch (0.46 m) telescope at Mount Palomar to take photographs of sections of the sky for seven days each month, and they asked the younger whiz to join them. It would prove to be an immensely successful collaboration, even before the discovery of Shoemaker-Levy 9.

Neither Gene nor Carolyn Shoemaker were astronomers by training, either. Eugene M. Shoemaker was born in Los Angeles in 1928, and graduated from the California Institute of Technology at the young age of nineteen. He gained a master's degree in geology a year later, and was hired by the U.S. Geological Survey. Assigned to search for uranium deposits in several western states, he became increasingly interested in volcanoes and asteroid craters, both those he encountered during his work (uranium is often found in the vents of extinct volcanoes) and those on the moon. Gene earned his Ph.D. from Princeton University in 1960, with a thesis on Meteor Crater, Arizona. As one of the pioneers of planetary geology, and the effect of asteroid impacts on Earth, he was in the right place at the right time to become NASA's chief consultant on the subject, training the Apollo astronauts as they prepared to journey to the moon. He had long wanted to go to the vastly cratered surface of the moon himself, but was diagnosed with Addison's Disease (which the world would later learn also affected President John F. Kennedy), which ruled out any such journey. In 1969, while retaining his ties to the U.S. Geological Survey, he became a professor at Caltech. At this point he became interested in extending his study of craters to the objects that created them—asteroids and comets. By 1973, he and colleague Eleanor Helin, assisted by students, were looking for asteroids and comets using the small Schmidt telescope at Mount Palomar.

Shoemaker had married Carolyn Spellman in 1951, and once their two children were grown, she joined him in his work at Mount Palomar, starting in 1980. Although her own college degrees were in history and political science, she proved to have a great gift for analyzing the photographs taken with the Schmidt telescope. By the early 1990s, she had discovered twenty-seven comets and more than 300 asteroids, a tenth of which were in orbits that brought them close to Earth. Thus it was hardly surprising that the Shoemakers, essentially self-taught astronomers themselves, should get along so well with David Levy, who shared their fascination with comets and had developed, like Carolyn Shoemaker, a remarkable instinct for spotting them.

All astronomers, professional or amateur, need good instincts. They must be able to pick out a tiny point of light that somehow seems different or out of place in a vast sea of such lights. There are, after all, some 200 million suns in our own small corner of the universe, the Milky Way. Comets can wink in and out of view from Earth in a matter of days—only large comets whose orbits carry them close to Earth are visible for long, even through telescopes. The comet hunter must know the sky, at all seasons, like the back of his or her own hand to be able to spot the minuscule anomaly that might be a comet. Only a few are discovered each year, as few as seven, sometimes as many as a dozen. Moreover, a comet that may seem new can turn out to be one of the 160 that make periodic returns—keep

in mind that it took David Levy nineteen years to find the first
new comet that would bear his name.

Once the Shoemakers and Levy teamed up, their combined
expertise began to pay quick dividends, as demonstrated by the
following list of their photographic discoveries at Palomar:

- Periodic Comet Shoemaker-Levy 1, 1990o
- Periodic Comet Shoemaker-Levy 2, 1990p
- Comet Shoemaker-Levy, 1991d
- Periodic Comet Shoemaker-Levy 3, 1991e
- Periodic Comet Shoemaker-Levy 4, 1991f
- Periodic Comet Shoemaker-Levy 5, 1991z
- Comet Shoemaker-Levy, 1991a1
- Periodic Comet Shoemaker-Levy 6, 1991b1
- Periodic Comet Shoemaker-Levy 7, 1991d1
- Periodic Comet-Shoemaker-Levy 8, 1992f
- Periodic Comet Shoemaker-Levy 9, 1993e
- Comet Shoemaker-Levy, 1993h
- Comet Shoemaker-Levy, 1994d

This official list, and that of David Levy's backyard discoveries,
are drawn from his formal résumé at www.jarnac.org.

As part of the Shoemaker team, Levy also discovered, together with Henry Holt, in June 1990, Asteroid 5261 Eureka, the
first Martian Trojan asteroid. It is called a Trojan asteroid because it shares the orbit of Mars itself.

This list would be impressive for a short period of time even
if it did not include Shoemaker-Levy 9, but it was that comet
that thrust Levy into the national spotlight in an entirely new
way. He had published eight books on astronomy, starting in
1982, including the widely read *The Sky: A User's Guide,* first published by Cambridge University Press in 1991. But the publicity
surrounding the crash of the comet into Jupiter would bring hundreds of interviews with reporters for magazines, newspapers,

radio, and television. Gene Shoemaker had experience at handling this kind of pressure from his appearances with Walter Cronkite on CBS during coverage of the Apollo moon landings, but for Levy and for Carolyn Shoemaker, it was a new world. By the time it was over, all three found themselves close to exhaustion.

On the day they discovered Shoemaker-Levy 9, however, they had no inkling of the hype that would eventually develop. In fact, they weren't quite sure what they were looking at to start with. As Levy related in his subsequent book, *Impact Jupiter: The Crash of Comet Shoemaker-Levy 9,* the team was on the second night of one of its monthly seven-day sessions at Mount Palomar. The sky was somewhat overcast, and they considered not trying to get any photographs at all. But Levy didn't want to lose an entire night, and the Shoemakers agreed to give it a try. When Levy loaded the film into the telescope, he noticed through the eyepiece that Jupiter was glowing brightly in the middle of the field they were supposed to photograph. (Astronomers search the sky in mathematically discrete fields in order to keep precise track of what they are looking at.) The brightness of Jupiter would obscure some objects in the field, but they decided to go ahead anyway. Three eight-minute exposures were taken, and the team then waited the requisite minimum of forty-five minutes to take a second set of exposures. The wait between exposures was crucial to determining if any object had moved, which would suggest either a comet or an asteroid. But in this case, because of cloud cover, there was a longer than usual delay between taking the two sets of exposures, which meant that a comet might move so much that it would be difficult to be certain it was the same object.

To view developed photographs, the team used a device Gene Shoemaker had put together a decade earlier for his work as a geologist, a stereomicroscope that gave an impression of three-dimensional depth when used on pictures of geological terrain. It had also proved invaluable in comet hunting, since a moving object would seem to float against the background of other objects. The developed photos of the area around Jupiter that had been taken on the night of March 23, 1993, were finally examined by Carolyn Shoemaker on the afternoon of March 25. Her husband and Levy were there with her, busying themselves with other things, but ready to consult with her if she spotted anything. What she found was very peculiar. She had initially passed by something that gave the impression of "a floating edge-on galaxy." But only objects in our own solar system are close enough to produce the floating effect, and she went back for a second look. It was a bar of light with a coma (the hazy cloud of gas and dust that surrounds the nucleus of a comet) as well as several tails.

Several tails? That didn't make sense, either, unless it was what Carolyn Shoemaker would call a "squashed comet." Gene Shoemaker and then David Levy, as well as a guest, Phillipe Bendjoya, an expert on asteroids, took a look. Peculiar as the object was, it did give the appearance of some sort of comet. The team immediately sent a message to Brian Marsden, the director of the International Astronomical Union's Central Bureau for Astronomical Telegrams in Cambridge, Massachusetts. Sightings from all over the world were kept track of there, making it possible to know with certainty who had first discovered a new comet, asteroid, or other celestial body. Marsden e-mailed them back with the suggestion that an Australian observer, Rob McNaught, take a look, and it was quickly agreed that Jim Scotti, whom Levy knew was observing that night with the thirty-five-inch (0.89 m) Spacewatch camera at Kitt Peak, should also be asked if he could confirm the sighting. As the team drove back to their lodgings to have dinner, Gene Shoemaker suggested that the object might not just look as though it were close to Jupiter, but in fact really be so close that Jupiter's gravity had broken it up.

The possibility remained that it wasn't even real. Jim Scotti was initially skeptical, thinking it might just be a reflection on film, since it was moving in the same direction and at the same speed as Jupiter. Refusing to be discouraged, the team decided to go over to the main dome at Palomar, which houses the forty-eight-inch (1.2 m) telescope, to get astronomer Jean Muller's assistance in measuring the object's orbit, using ten stars as reference points. It was indeed very close to Jupiter. When that task was finished two hours later, they returned to the smaller dome and Levy called Jim Scotti. His voice sounded strange, and he said he was "picking his jaw up off the floor." He confirmed that not only did they have a comet, but a highly unusual one. It appeared to have at least five nuclei, traveling side by side, with wings on each side. He thought there were probably more nuclei, but would have to wait until the sky cleared to take another look.

With Scotti's confirmation and additional measurements, Brian Marsden in Cambridge issued a discovery circular with complete technical details as far as they were then known. The team knew what it had. Carolyn Shoemaker, who had already become second on the list of all-time comet discoverers (behind an eighteenth-century French astronomer) said that this comet was the most exciting discovery of her life. As for David Levy, the Shoemakers teased that they had to pull him down off the ceiling.

———

Just over a week after the discovery of Shoemaker-Levy 9, Brian Marsden had refined the calculations about its orbit that he had included in the original circular. It now seemed clear that it was in orbit around Jupiter, had come very close to the planet in 1992, and could be expected to return. Within two months, as calculations about the orbit came in from around the world, it became evident that the 1992 approach to the planet had been within the so-called Roche limit. This limit is named for the nineteenth-century French mathematician Édouard Roche, who calculated the point at which a smaller body would be torn apart by the gravity of a planet. That explained why the comet was in so many pieces. Then Marsden issued a new circular. As Levy points out, such circulars are not given journalistic headlines, no matter how startling the news they contain. But any astronomer would immediately see that the calculated orbit meant the pieces of the comet would collide with Jupiter on the next pass, which was expected to take place the following July. For a long time—no one was certain how long—Shoemaker-Levy 9 had been drawn into an ever tighter orbit around the giant planet. The next circuit would be its last, ending with impact.

Astronomy is an extremely varied field. While all astronomers will be interested in a major development or discovery in any aspect of the discipline, there is little technical connection between searching for comets and searching for invisible black holes in other galaxies. But sometimes there is an unusual conjunction of interests, and Shoemaker-Levy 9 provided the occasion for just such an overlap. The collision with Jupiter would provide a great deal of new information to comet specalists as to what happens when a comet crashes into a planet. To experts on Jupiter itself, the collision might reveal a great deal about the mysteries of the planet's interior. Impact physicists would have the chance to study the results of the vast releases of energy— "hundreds or thousands of times stronger," Levy writes, "than all the world's nuclear arsenal put together."

Excitement about the collision was somewhat dampened when it became clear that the impacts of the fragments would take place on the dark side of Jupiter, out of view of Earth and the Hubble telescope. The *Galileo* spacecraft, launched in October 1989, was en route to Jupiter, and would still be seventeen months from the planet in July 1994. Fortunately, the angle of its approach would allow a view of part of the dark side of Jupiter, and its camera ought to be able to record at least some of the now expected twenty-one separate impacts, taking place over a week's time. Many problems remained to be solved, however, since *Galileo* would have to be preprogrammed to take any pictures, and there was still great uncertainty about the timing of the impacts. The main antenna of *Galileo* had not opened as it was supposed to, which meant that sending data back to Earth was extremely slow. There was also concern that the light released by the impacts might damage *Galileo* in ways that would compromise its main Jupiter mission nearly a year and a half later.

Quite aside from the technology involved, two different ways of identifying the twenty-one fragments had evolved. One used letters, the other numbers, and they reversed one another directionally, with the letters moving from east to west, and the numbers from west to east. That meant that a single fragment would have two designations, such as Q and 7. The media eventually solved that problem by ignoring the numbers. The letter system made more sense even to many scientists, since it had been used to designate the pieces of Periodic Comet Brooks 2 as far back as 1889.

There was a great deal of debate about the actual size of the fragments, which would affect the force of the impacts and the amount of energy released as they struck the planet. Knowledge about the lower layers of Jupiter's strange atmosphere and its supposed metallic core was limited, largely a matter of speculation based on computer models, which was exactly why *Galileo* had been dispatched to study it. There were thus three scenarios about what might happen, with one suggesting impact flares so huge and so bright that they might be seen from Earth without a telescope, while the other two held out far less hope of a great visual show, either because the comet fragments would disintegrate too high in Jupiter's atmosphere, or would be dragged into such depths that little evidence of disruption would appear on the surface.

The range of possibilities had David Levy extremely excited one week and quite discouraged another, and the Shoemakers went through similar ups and downs. Conflicting information continued to pour in, including Hubble data in September 1993 and March 1994 indicating that the fragments Q and P had split apart into smaller units. Did that mean that the whole train of comet fragments might disintegrate even before impact? The media began to downplay the possibilities of any spectacle, then reversed course as the July 1994 impact date approached. The major magazines, television networks from CNN to the BBC, and, of course, science magazines of every stripe all wanted interviews. Levy had two new books coming out, *The Quest for Comets* and *Skywatching*, and began a cross-country lecture tour at the beginning of April that had burgeoned into nearly a hundred talks, as well as book signings. On May 17, Shoemaker-Levy 9 showed up on the cover of *Time*. The hype was on, and some scientists tried to cool down public expectations by emphasizing the possibility of a complete fizzle. The Shoemakers and Levy were almost relieved when the arrest of O. J. Simpson for the murder of his former wife grabbed the headlines in mid-June. Perhaps they'd get a bit of breather before the collision a month later.

The experience David Levy was undergoing was an unusual one for an amateur scientist. It lies at the opposite extreme from the disinterest and neglect that attended Gregor Mendel's breakthrough discoveries about genetics. Mendel's experiments with peas consumed nine years of painstaking work and original thinking, but the two lectures he eventually gave on the subject were heard only by a small group of local dignitaries, and the published paper containing those lectures was largely ignored for thirty-five years. David Levy, on the other hand, had a vast national, and often international, audience paying close attention to what he (and the Shoemakers) had to say about the comet they had discovered. Part of the difference, of course, is due to modern communications, particularly television, which could not only bring Levy live to a far-flung public, but also provide computer graphics that made the upcoming collision between the comet and the planet Jupiter vividly immediate. More crucially, Mendel was promulgating a new scientific *concept*, one that involved invisible processes, while Levy had discovered a new *object*, easily seen with a telescope. Throughout the history of science, new concepts have often been ignored or have met with active resistance by more established scientists. Charles Darwin's theory of evolution was bitterly challenged, while Albert Einstein's first papers in 1905 were initially understood by almost no one except Max Planck. A celestial body that can be photographed, like Shoemaker-Levy 9, doesn't run into such problems.

There are also more subtle differences, and similarities, to be considered. Gregor Mendel's "laboratory" was a monastery garden and greenhouse—not all that different a setting from the backyards in Canada and Arizona where Levy spent so many years peering at the night sky through small telescopes. Both men were curious about the natural world. There is some evidence that Mendel knew how important his work might prove, but in Levy's case it was more a matter of intellectual curiosity and personal satisfaction. To the extent that Levy expected to have an impact on the scientific world, it involved teaching children about astronomy. For two decades before the discovery of Shoemaker-Levy 9, he had been introducing groups of children to the wonders of the night sky, working with schools and local associations. He was very good at this kind of instruction, conveying his own enthusiasm with ease and humor and knowing just how to phrase things so that young minds could grasp them. As the years passed, the fact that he himself had discovered a number of comets gave him great credibility with the children (and occasional groups of adults) whom he tried to turn on to the pleasures of astronomy. It was true that his discoveries had gradually gained him a serious reputation among both amateur and professional astronomers as a man with special gifts. But while he was a respected figure in the fairly tight-knit world of astronomy, and through his books was succeeding in reaching a wider audience, he was not in it for fame. Every amateur astronomer no doubt harbors a few fantasies about making a remarkable discovery, but they also know that it involves luck as much as anything else.

Luck certainly played a considerable role in the discovery of Shoemaker-Levy 9. The comet had, after all, broken up on its previous journey around Jupiter, and someone else might have noticed what was going on then. In fact, there was another group, also working at the Palomar locale, that had taken pictures of the same part of the sky a week earlier. The comet was there, but it hadn't registered with this group, perhaps because its features were so odd as to suggest a mere flaw in the photos. On the night of March 23, 1993, when the crucial pictures had been taken by the Shoemaker-Levy team, the weather was so overcast that there was some discussion about whether the group should even bother to do anything. Levy's habitual enthusiasm had been the factor that led the group to give it a try. There had also been a problem with the film to be used that night. Someone had opened the box of prepared film, spoiling the negatives. But Gene Shoemaker thought some of the film lower in the box might be all right, and it was. It is also worth emphasizing again that when Carolyn Shoemaker was going over the developed film two days later, she passed by the odd configuration once, and then went back. She could not quite believe what she was seeing, but both her husband and David Levy had agreed that they might indeed be looking at a "squashed comet." They had then proceeded to report it on what was largely a hunch. The group that had ignored it—or not seen it at all the previous week—had been made of up of professional astronomers. Professionals get hunches, too, of course, but they may be less open to them than amateurs who have less to lose if they make a mistake.

Obviously, astronomy is a field that is particularly congenial for amateurs. You have to know what you're doing to discover a new comet or asteroid, but it does not require an advanced degree. Nor are there great expenses involved. A modest backyard telescope is all that is needed to make your mark. If you become a successful amateur, access to a larger telescope like the eighteen-inch (0.46 m) one at Palomar can often be arranged. What's more, professional astronomers these days are likely to be focused on much larger questions, like the search for planets orbiting other stars. Such work takes a great deal of professional training, and the most advanced technology—it is not work for amateurs. Because the cutting edge of astronomy—the kind of work that can bring great professional rewards, perhaps even a Nobel Prize in Physics—requires so much time, training, and effort, activities like looking for new comets are largely left to amateurs. If a professional spots one, he or she will be certain to report it—it's always a feather in your cap to have your name attached to a celestial object. But on the whole, the professionals have bigger fish to fry, which means that amateurs have an opportunity to make a name for themselves in astronomy somewhat more easily than is the case in most scientific fields.

Nevertheless, it remains very unusual for an amateur astronomer to find himself or herself suddenly accorded the kind of media attention that David Levy was getting in the spring of 1994. He has commented that he came to much more fully understand the complaints that people constantly in the public eye often make about the fishbowl lives they lead. Ironically, even as the attention was peaking, Levy and the Shoemakers were concerned that the media attention might suddenly dry up because of something else that was happening. The O. J. Simpson case was just one example of what could occur to push the comet's crash into Jupiter into the background. Two days before the impact of the first fragment, there were rumors in the press that the United States was on the verge of launching a military invasion of the troubled Caribbean nation of Haiti, as had occurred with Grenada during the Reagan administration. Such an event would utterly eclipse the media's interest in a comet colliding with a planet hundreds of millions of miles from Earth.

There was also a concern that expectations had been raised too high. If the comet fragments turned out to produce little effect on Jupiter, it could undermine the greatly increased public interest in astronomy that had been growing over the past several months. David Levy, in particular, worried that a fizzle would have a severely negative effect. The passage of Comet Kohoutek in 1973 had been greatly hyped—there were estimates that it would be as bright as the moon—and cruise lines even booked special "Kohoutek" voyages. Because of a variety of factors, it turned out to be a dud of a comet as far as viewing it from Earth was concerned. Comet Shoemaker-Levy 9 had the potential to interest an entire new generation in astronomy, but if it fizzled, a backlash could easily set in. Having spent his entire adult life celebrating the joys of amateur astronomy, Levy would have been devastated if this remarkable comet bearing his name failed to put on the expected show.

But luck held. Haiti was not invaded. Indeed, Levy was astonished to find his picture on the front page of USA Today on the morning of July 16. It had also now been determined that most of the fragments would crash into Jupiter in full view of Galileo's camera, which meant that the moment of impact on the dark side of Jupiter could be recorded. What's more, the rotation of the planet would carry the impact areas into full view of the Hubble Space Telescope within twenty minutes of impact, instead of two hours. Thus, if the impact did seriously disrupt the atmosphere of the planet, it should be possible to view the results before they could dissipate.

David Levy would have liked to watch through a telescope when impact occurred, but instead he and the Shoemakers were staying in Washington, D.C., where they would have access to the Naval Observatory on the grounds where the official Vice Presidential residence is also located. They would spend most of their time dealing with reporters, however, in an auditorium of the Space Telescope Science Institute in Baltimore, Maryland, where the first pictures from the Hubble would come in. The first fragment to crash into Jupiter, A, was one of the smaller ones, which meant that the possibility of disappointment was large. The impacts would continue for a week, but if A didn't show at least some effects, Levy knew the press and the public could lose interest very quickly.

The first word that the crash of fragment A had produced a plume (an upward geyser of atmospheric gases) on Jupiter came from a small observatory in Spain. The plume was almost immediately confirmed from Chile. Reporters were already excited by that news when Levy and the Shoemakers entered the hall for the third news conference of the day at 7:30 P.M., but it would be another two hours before the first pictures would be transmitted from the Hubble, a delay caused by the fact that a space shuttle was in orbit—its communications with Earth had priority over data from the Hubble. When they did arrive, those pictures showed a plume shooting off the edge of Jupiter. Even little A was having a major effect. All around the world, astronomers gasped. Nothing like this had ever been seen before. A planet, and not just any planet, but the largest in the solar system, was being deeply wounded by an object from space, an object perhaps no more than a quarter of a mile across (there are still arguments about size), but one that was hitting the planet at a minimum speed of 60,000 miles per hour.

In the course of the following week, the fragments of Shoemaker-Levy 9 crashed into Jupiter one by one. Three of the small fragments were fizzles, probably because they contained less solid material than had been surmised from their brightness. But the rest sent plumes 1,800 miles (3,000 km) above Jupiter's atmosphere, and the fireballs that flared for as long as ninety seconds left dark spots in a curving line along the bottom quadrant of the planet. When the fragmented comet had first been viewed, it had been called a "string of pearls." Now, Carolyn Shoemaker would say, it had become "a necklace of garnets." The biggest impacts, of fragments G, H, K, and L, left such large spots that almost anyone with a backyard telescope could see them. The sale of small telescopes, it was reported, rose sharply. On the fledgling Internet, the photographs of the impacts posted by NASA received more than a million hits in the course of the week. Television news carried daily updates, and Levy and the Shoemakers gave so many interviews that exhaustion was setting in. The entire world was getting a crash course in comets and planetary science.

Both the Hubble Space Telescope and the *Galileo* spacecraft did their jobs superbly, providing breathtaking pictures of the weeklong series of impacts. Taken together with the information pouring in from Earth-based telescopes, so much data was gathered that it would takes years to fully analyze. In the aftermath, Jupiter's atmosphere darkened to a degree that put to rest any remaining doubts about the effect the asteroid that created Earth's own Chicxulub crater would have had on the biosystem in which the dinosaurs lived 65 million years ago. Quite aside from the scientific bonanza it provided, the collision between Shoemaker-Levy 9 and Jupiter made scientists, politicians, and the public take more seriously the dangers to Earth itself posed by the comets and asteroids journeying through the solar system. Gene Shoemaker had been trying for decades to persuade the scientific community to think about contingency plans against the possibility that a future comet or asteroid might collide with Earth. Finally, that danger was taken seriously.

Almost exactly three years after Shoemaker-Levy 9 plowed into Jupiter, Gene Shoemaker was killed in an automobile accident in Australia. His wife was also seriously injured, but recovered. As soon as the news of his tragic death was announced,

Carolyn Porco, a planetary scientist who had been one of his students, suggested that his ashes be carried to the moon on the *Lunar Prospector* spacecraft, which was due to be launched in January 1998. Carolyn Shoemaker and David Levy both enthusiastically supported the idea, and NASA agreed with alacrity. A tiny polycarbonate capsule was imbedded in a vacuum-sealed aluminum sleeve deep inside the *Prospector*. As NASA would report in a press release: "Around the capsule is wrapped a piece of brass foil inscribed . . . with a passage from William Shakespeare's enduring love story, *Romeo and Juliet*":

> And when he shall die,
> Take him and cut him out in little stars,
> And he will make the face of heaven so fine
> That all the world will be in love with night,
> And pay no worship to the garish sun.

David Levy has produced another ten books since the momentous days of 1994, including a biography of Gene Shoemaker called *Shoemaker by Levy: The Man Who Made an Impact*. There have been several books directed to children and young adults. Levy has used the fame he gained from Comet Shoemaker-Levy 9 to further what he always regarded as his primary mission: to inspire adults, but particularly young people, to look up at the night sky, to watch, to learn, and to take joy in the great panorama arching above them.

Levy did not become a comet hunter for anything but the amateur's joy of learning and growing. Yet, more than most amateurs, he succeeded not only in inspiring others, but in becoming a noted expert in a field he never formally studied. Back in July 1994, in the heady atmosphere of excitement about the collision between Comet Shoemaker-Levy 9 and Jupiter, Levy had arrived in Baltimore for the suspenseful first night of impact accompanied by his mother, Dr. Edith Levy, Steve O'Meara of *Sky and Telescope*, and George Cruys of The Nature Company.

As Levy relates in *Impact Jupiter*, a guard at the institute checked their names on a list of dignitaries. O'Meara, Cruys, and Levy's mother were duly checked off, but David Levy's name was not there. The guard was not impressed that the comet carried his name—he wasn't on the list, period. The group finally persuaded the guard to radio inside. A disembodied voice chuckled and replied, "He's too high on the list to be on the list. Let him through."

Not bad for an amateur.

To Investigate Further

Levy, David H. *Impact Jupiter: The Crash of Comet Shoemaker-Levy 9*. New York: Plenum Press, 1995. This inside account of the discovery of the great comet and its collision with Jupiter is, quite naturally, definitive. Levy is very good at getting across complex scientific material in an easily digestible way, he has a delightful sense of humor, and he fully conveys the excitement and suspense created by this extraordinary cosmic event.

Levy, David H. *Comets: Creators and Destroyers*. New York: Simon & Schuster, 1998. A modestly priced trade paperback that deals with all aspects of comets, from their formation to the possibility of one colliding with Earth. Lively, authoritative, and well written.

Sagan, Carl, with Ann Druyan. *Comet*. New York: Ballantine, 1997. A revised and updated version of one of Carl Sagan's most highly regarded books, lavishly illustrated with more than 300 photographs and commissioned paintings.

Beebe, Reta. *Jupiter: The Giant Planet*. Washington, D.C.: Smithsonian Institution Press, 1996. Beebe was part of the Hubble Space Telescope team observing the collision between Shoemaker-Levy 9 and Jupiter. Her updated book is intended for the general reader, but although it is clearly written, some sections are necessarily quite technical.

Levy, David H. *Shoemaker by Levy*. Princeton, NJ: Princeton University Press, 2000. This tribute to Levy's friend and fellow comet discoverer does full justice to a remarkable scientist.

Harland, David M. *Jupiter Odyssey: The Story of NASA's Galileo Mission*. New York: Springer-Verlag, 2000. *Galileo's* pictures of the 1994 collision were extremely valuable to scientists, and that material is well covered here. But *Galileo's* long voyage to Jupiter is a much bigger story, and this book tells it fully and excitingly.

Levy, David H. *Skywatching*. New York: Time-Life Custom Publishing, 2000. If you are interested in becoming a backyard astronomer, this is the book to get, written with Levy's usual enthusiasm and attention to detail.

Note: The Comet Shoemaker-Levy Home Page (www.jpl.nasa.gov/sl9) is one of NASA's most popular sites, visited nearly eight million times. It has a wide range of links to other web sites, including more than 1400 images from 64 observatories.

CHAPTER 3

Henrietta Swan Leavitt

Cepheid Star Decoder

The women worked in a small room at the Harvard College Observatory on Garden Street in Cambridge, Massachusetts. Several heavy wooden desks had been fitted into the room, at angles to one another. The long, prim Victorian skirts the women wore hardly made maneuvering in these congested quarters any easier. But the women, known as the "computers," did not complain. Nor did they object to the pay, which started at a mere twenty-five cents an hour. The work they were doing, sorting thousands of photographic plates of the stars, was tedious in many ways, but it could also be exciting. These women were in a sense privileged for that period—they were working in science, unusual in itself for a woman, and astronomy was a field that was becoming more important with every passing year. They were pioneers, in more ways than one, and they knew it.

Astronomy had taken a great leap forward in the early seventeenth century when Galileo put together the first true telescope. Nevertheless, the view through early telescopes was fuzzy around the edges. Isaac Newton fixed that by understanding that refracted light, using mirrors, would produce a sharper image. Throughout the eighteenth and nineteenth centuries, telescopes became larger and more complicated, but what was seen through them could be conveyed to a broader public only through words and drawings. Many of the drawings were beautiful, but they seldom reflected with great accuracy the mathematics of astronomy, tending to distort the perfect equations that Newton had introduced with his laws of motion and gravity.

The invention of photography changed all that. The earliest photographs in the collection of the Harvard College Observatory were either daguerreotypes or collodion (wet) plates taken between 1849 and 1885. By the 1880s, technical improvements had made photographic plates a basic tool of astronomy. Yet a problem had arisen that exists even today. Photographs of the moon, the planets, and the stars can be amassed at a rate greater than the amount of time astronomers have to examine and investigate them properly. Full analysis of the photographs taken by such spacecraft as *Voyager* or *Galileo*, for example, always lags several years behind their transmittal to Earth. Before photography, astronomers had too little evidence on which to base their theories of the universe; once photography was fully established, it was difficult to keep up with the evidence.

Edward C. Pickering, the director of the Harvard College Observatory, decided that a massive organizational effort was necessary, and starting in 1886, he began hiring women specifically to catalogue the thousands of plates that were piling up. Almost all the women Pickering hired, some married women and some spinsters, had college degrees, and a few had taken courses in astronomy. Pickering was very much a Victorian gentleman, and in his view women were too delicate to be put to work in the cold telescope dome in Cambridge, let alone sent to Peru, where another telescope was located in the mountains, at Arequipa. He was also aware that he could hire women to do the cataloging at much lower pay than he could young men. In one sense, Pickering's attitude can seem infuriating today. He was using society's prejudices against women to save a buck—a problem that still remains with us in too many fields. At the same time, however, he gave to these bright—in some cases truly brilliant—women an opportunity that could not be found anywhere else until later. And several of the "computers" made the most of that opportunity, in ways that would sometimes astonish even Pickering.

The three women who left the greatest mark on astronomy were Williamina Paton Stevens Fleming, Annie Jump Cannon, and Henrietta Swan Leavitt. The first of three to become a member of the observatory staff was Fleming. Born in Scotland in 1857, she married at the age of twenty and emigrated to America. But when her marriage collapsed in 1879, she took a job as housekeeper with Edward Pickering, who soon recognized her intelligence (she had started teaching younger classmates in Scotland when she was only fourteen). By 1881 she was a full-time member of the observatory research staff. She developed a technique for classifying the spectra of stars (the line patterns caused when a star's light is dispersed by placing a prism in front of a telescope lens). It is known as Pickering-Fleming system; Pickering's name went on almost everything the "computers" did. Fleming analyzed tens of thousands of astronomical plates for what would eventually be published as the Henry Draper Catalogue, in honor of a wealthy Harvard graduate and amateur astronomer who subsidized the work. In the process, she discovered 10 novae, 52 nebulae, and 220 of the variable stars called Cepheids. She was appointed curator of the observatory's photographic collection in 1888, and published several works on astronomy before her death in 1911.

Fleming's successor as curator of the collection was Annie Jump Cannon, born in Delaware in 1863. Cannon graduated from Wellesley College in 1884, and spent the next decade traveling, taking photographs, and studying music. In 1895, she became serious about astronomy, taking an advanced study year at Wellesley, and then transferring to Radcliffe expressly to become a student of Pickering. He hired her as an assistant at the observatory in 1896, and she joined forces with Fleming to work on the Henry Draper Catalogue, which was eventually published in nine volumes from 1918 to 1924. She would classify the spectra of 225,000 stars, and discover 300 variable stars and 5 novae. She was appointed a professor of astronomy at Harvard in 1938, three years before her death. Perhaps more significantly, her work earned her the first honorary doctorate ever given to a woman by Oxford University, awarded in 1925.

Both Fleming and Cannon hold a distinguished place in the history of American astronomy, and Cannon was rewarded with a number of academic honors. Yet it was Henrietta Swan Leavitt who made the most important contribution of the three women. She discovered something that provided the key to an entirely new understanding of the universe. Like Fleming and Cannon, Leavitt became one of Pickering's "computers" by a somewhat roundabout route. She had been born near Cambridge in Lancaster, Massachusetts, the daughter of a Congregational minister, on July 4, 1868. From 1886 to 1888 she attended Oberlin College in Ohio, and then transferred to Radcliffe, which was known at the time as the Society for the Collegiate Instruction of Women, but was already establishing a reputation for excellence based on the fact that it drew heavily on the Harvard faculty. In her senior year she became interested in astronomy, and took an additional course in the subject following her graduation. She then fell seriously ill. The exact nature of the illness is not known, since she never talked about it, but she was left with a serious hearing loss. She may have had scarlet fever, a childhood disease that sometimes struck young adults and could cause deafness in severe cases.

Despite her handicap, she offered herself as a volunteer worker at the Harvard College Observatory in 1895. As it happened, Edward Pickering was teaching Annie Jump Cannon that same year, and would give her a job a year later. Since Cannon had also developed a less severe hearing impairment, it may well

have seemed to him that the two women would be able to bol-
ster one another. Leavitt worked as a volunteer for several years,
and then was given a permanent place on the staff in 1902. Like
Fleming and Cannon, Leavitt worked on the Draper project. As
they tried to determine the magnitude (brightness) of all measurable stars, it became clear to Pickering that Leavitt had a
particularly good eye, and soon after she became a full-time
employee, he named her head of the photometry department.

The brightness of stars was one thing, their size quite
another. Among astronomers, there were increasing debates
about whether brightness was an indication of size or of distance from Earth. At the time, it was assumed that all the stars
visible from Earth were part of a single galaxy, our own Milky
Way, and that the Milky Way comprised the entire universe. In
addition, it was believed that our own solar system must be
quite close to the center of the Milky Way, since stars seemed to
be evenly distributed in all directions. No matter where on Earth
an astronomer peered through a telescope, there were vast numbers of stars to be seen. The fact that different stars were seen
from the southern hemisphere seemed to corroborate that Earth
must be near the center of the universe. If it were not, surely
the number of stars would thin out in one direction or another.
From a twenty-first-century perspective, this can seem both
naive and arrogant, a holdover from the Ptolemaic view that the
sun revolved around Earth. Somehow, we humans had to be at
the center of everything.

We have become so used to revolutions in science that it is
easy to lose sight of how slowly new ideas were developed in
previous centuries. Ptolemy's model of the universe, with Earth
at its center, and the moon, other planets, and the sun revolving
around it in perfect circles, was promulgated in the second century A.D. He believed that the stars were somehow attached to a
vast dome that surrounded the whole shebang—although it also
seemed possible that the stars might be shining through holes

pricked in the dome, an idea later taken to suggest that the light beyond the dome was that of the Christian Heaven. Even Ptolemy (90–168) recognized that the orbits of Venus, Mercury, and Mars weren't properly aligned with that of Earth. He got around this difficulty by suggesting that they revolved around a point beyond Earth that in turn revolved around our home planet. This awkward theory managed to prevail for 1,400 years. It satisfied the Roman Catholic Church because it seemed to confirm the idea of masterful Creator (whose realm was clearly beyond the dome of stars), and because even early scientists who should have known better liked to think of themselves as being at the center of the universe.

Nicolaus Copernicus spent thirty years at the beginning of the sixteenth century trying to work out a new model. He took into account the movement of Earth on its own axis, explaining that the movements of Venus and Mercury were more limited because their orbits were closer to the sun than that of Earth, and that the known outer planets, Mars, Jupiter, and Saturn, had peculiar orbits because they were moving more slowly than Earth around the sun. But he wasn't able to free his theory from the concept of perfect circles, which meant a complicated system involving Earth revolving around an invisible center that revolved around another invisible center, with that point in the sky revolving around the sun. It took a series of observations, discoveries, and new mathematical formulas devised by several scientists to begin to get matters straight. The close observations of the planets made without a telescope by the Danish astronomer Tycho Brahe (1546–1601) led the German astronomer and mathematician Johannes Kepler (1571–1630) to develop his three laws of planetary motion, which at last got the planets into proper elliptical orbits. Isaac Newton's 1687 laws of gravity and motion finally explained how it all actually worked. Since these developments had taken 1,500 years, the fact that most astronomers still located Earth close to the center of a single galaxy a mere 200 years later is not all that surprising.

There had been hints, however, that the universe was more complex, and vastly larger, than the available evidence suggested. In the late eighteenth century, thanks in part to the predicted reappearance of Halley's Comet in 1758, comet-watching was all the rage among astronomers. Unfortunately, the night sky was full of misleading objects, such as the hazy patch in the constellation Andromeda that can be seen even with the naked eye. As physicist Steven Weinberg explains in *The First Three Minutes*, "In order to provide a convenient list of objects *not* to look at while hunting for comets, Charles Messier in 1781 published a celebrated catalog, *Nebulae and Star Clusters*. Astronomers still refer to the 103 objects in this catalog by their Messier numbers—thus the Andromeda Nebula is M31, the Crab Nebula is M1, and so on." Some astronomers and theorists suggested that these fuzzy patches might actually consist of stars in other galaxies that were rendered hazy by enormous distances. Weinberg believes that the German metaphysical philosopher Immanuel Kant was the first to broach this idea, way back in 1775. But two problems stood in the way of proving this theory. First, no telescope existed with enough power to reveal that the fuzzy patches did consist of individual stars. Second, those who held that Messier's nebulae and star clusters lay within our own galaxy were mostly correct—two-thirds of them, it would eventually turn out, are part of the Milky Way. Caution on the subject of multiple galaxies was entirely justified.

In 1904, when Henrietta Swan Leavitt began her most important work, there were still no telescopes capable of revealing the individual stars within the Andromeda Nebula, and it would be another nineteen years before such a telescope was built. There were, however, hundreds of thousands of photographic plates of the stars being categorized by the ladies at the Harvard College Observatory. It was time for another major breakthrough in astronomy, and it would be Leavitt who supplied the means for measuring how far away the stars in the Andromeda Nebula really were, once a telescope with the power to reveal them came into existence.

———

Leavitt's chief research duty, so far as Edward Pickering was concerned, was his ambitious project to establish standardized values for the magnitudes of the thousands of stars captured on the photographic plates in the ever-growing observatory collection. She carried out this work with great diligence, starting with forty-six stars in the north polar region of the sky, and then adding more stars from the same area, eventually refining the standards of brightness down to the twenty-first magnitude. Over the next fifteen years, Leavitt used the system she developed to measure the magnitude of key stars in 108 regions of the sky, using plates from observatories around the world. These standards were invaluable to astronomers, and the system she devised remained in general use for several decades, until computer technology, which could produce far more accurate measurements, eventually replaced it. But although this project was a major accomplishment in itself, it was an entirely different achievement that would be her greatest contribution to the history of astronomy.

The telescope the Harvard Observatory maintained at the Boyden Station in Peru provided thousands of photographs of the Magellanic Clouds visible in the southern hemisphere. These had to be examined in the course of her regular work on magnitudes, but Leavitt became fascinated by a certain type of star that was somewhat more common in the Magellanic Clouds than in most regions of the sky. These were variable, or Cepheid, stars, named after the star Delta Cepheus. And thereby hangs a remarkable tale.

A few variable stars, whose brightness waxes and wanes over set periods ranging from as little as 5 hours to as much as 127 days, are visible to the naked eye, and were noted in antiquity, but no one really studied or tried to explain them until the late eighteenth century. A young man named John Goodricke, born into a wealthy and titled English family in 1764, became enthralled with astronomy in his teens. Goodricke was completely deaf, probably as the result of an early childhood fever, although he may have been born deaf. He was sent to the first known school for the deaf, in Edinburgh, Scotland, where he learned to speak and to read lips. It quickly became clear that he had a brilliant mind, and he was enrolled at Warrington Academy, a progressive Unitarian school where Joseph Priestley (see chapter 4) taught. Samuel Johnson was a tutor there at the time, and being somewhat hard of hearing himself, he took a special interest in Goodricke. It was probably another notable teacher, William Enfield, who first introduced him to astronomy.

At the age of seventeen, Goodricke returned home to live with his family in York, a short walk from the home of Nathaniel Pigott, who had built one of the three private observatories in England. Nathaniel's twenty-eight-year-old son, Edward, and Goodricke began observing stars together. Goodricke closely studied the double star Beta-Persei (also known as Algon-Winking-Demon). He calculated its periods of variability and wrote a paper, which he presented to the Royal Society in London, suggesting that the "wink" was due to the fact that one star in the double system was eclipsing the other, a theory that was confirmed more than a hundred years later. Even at the time, his paper created a scientific sensation, and after his observations and mathematics were confirmed by others, he was awarded the 1783 Geoffrey Copley Medal for the most significant scientific discovery of the previous year. He had just turned twenty.

Goodricke then turned his attention to Delta Cepheus, the fourth-brightest star in the constellation Cepheus, named for the mythological Greek king, whose vain wife Cassiopeia boasted that she and her daughter Andromeda were more beautiful than the Nereids who attended the sea god Poseidon. Poseidon took his revenge, flooding Cepheus's kingdom and demanding the sacrifice of Andromeda to a sea monster. Tied to a rock awaiting her fate, Andromeda was rescued by Perseus, who had just finished polishing off the dreaded Gorgon Medusa, whose stare turned humans to stone. He had been promised Andromeda's hand in marriage by her parents, but the wedding was interrupted by another suitor, a battle broke out, and Perseus found it necessary to use the head of Medusa to turn all the guests, as well as Andromeda's parents, to stone. Poseidon placed Cepheus and Cassiopeia among the stars, and they were eventually joined there by Perseus and Andromeda. All four constellations are visible in the same area of the northern sky in November.

Delta Cepheus, which forms one of the points of the king's crown, is, like Algon, a binary star, a yellow giant with a smaller blue companion. The yellow giant, as Goodricke was the first to discover, is also a variable star, but the eclipse solution Goodricke had proposed for Algon clearly was not applicable to the case of Delta Cepheus. Having been elected a member of the Royal Academy at the age of twenty-one, Goodricke continued his observations of Delta Cepheus. He caught pneumonia in the cold November air, and died just after his twenty-second birthday. While his paper on Algon established his reputation at the time, it was his discovery that Delta Cepheus was a variable star that would have the most lasting impact, since all stars of its odd type came to be called Cepheids.

Goodricke was better known at the end of the ninteenth century than he is now, and Henrietta Swan Leavitt would certainly have been aware of him and his work. Whether Goodricke's deafness, even more severe than her own, helped to inspire her to investigate variable stars, and particularly Cepheids, must remain a matter of speculation. But it is nevertheless remarkable that Cepheid stars, whose prototype had been discovered by a deaf young man more than a century earlier, should become the key to a major twentieth-century revolution in astronomy, a key provided by another hearing-impaired scientist.

Leavitt's work on variable stars in the Magellanic Clouds involved a time-consuming process of a kind that most people would find extremely tedious. She would find a negative of a photograph taken on one date, and lay it on top of a positive taken on another date. A star would show up as white on the positive but black on the negative. If the two tiny spots did not exactly coincide, it was a clue that the star might be a variable, brighter at some times than at others. That clue warranted further comparisons between other plates. In 1904 and 1905 alone, this technique led to the discovery of 1,054 variable stars in the Magellanic Clouds. Leavitt became ill again in 1907, but returned to work in 1908, the year in which she published a paper on the subject of variable stars in the *Annals of the Astronomical Library of Harvard*, listing a total of 1,777 variable stars. Among these, there were seventeen stars in the Small Magellanic Cloud that Leavitt studied in detail. She noted that these seventeen stars, whose cycles of brightness ranged from 1.25 to 127 days, differed in a highly significant way from one another: the longer the cycle, the greater the brightness. The implication of that finding was that Cepheids with similar cycles must be almost exactly alike in their properties. If that were the case, they could be used to measure celestial distances. If one star with a cycle that was the "twin" of another also was much brighter, it meant that the dimmer one must be farther away. Cepheid stars were mileposts across the visible universe.

No one paid any attention. In his biography of the great American astronomer for whom the floating space telescope is named, *Edwin Hubble: Mariner of the Nebulae*, Gale E. Christianson remarks, "Like Gregor Mendel's article on the hybridization of peas, Leavitt's paper was ignored." Obviously, Edward Pickering, who decided what would be published in the *Annals*, thought

Leavitt's findings were of sufficient interest to make them public. But neither he nor any other professional astronomer picked up on the real significance of what Leavitt had found. Leavitt herself was in no way a demonstrative person, and she certainly wasn't going to make a fuss with her boss. He had, after all, published her findings. Instead of jumping up and down and saying, "You don't understand," Leavitt simply amassed some more evidence. By this time, there were other areas of the sky to deal with in terms of Pickering's huge project on magnitudes. Yet Leavitt found the time to discover another eight Cepheid variables in the Small Magellanic Cloud that demonstrated the same relationship between the length of the cycle and the luminosity of the stars. In 1912 she published a second paper that can be found at this writing at www.physics.ucla.edu on the Internet. It is headed:

HARVARD COLLEGE OBSERVATORY
Circular 173
Edward C. Pickering, March 3, 1912
Periods of 25 Variable Stars in the Small Magellanic Cloud

Pickering noted, "The following statement regarding the periods of variable stars in the Small Magellanic Clouds has been prepared by Miss Leavitt." Leavitt included a table and two graphs in this presentation of her findings to make the significance of what she had found clearer. The article is only three pages long, and she duly notes that further study is needed. However, at the end of the second paragraph, she writes, "A remarkable relation between the brightness of these variables and the length of their periods will be noticed. In H.A. 60, No. 4 [a reference to her 1908 article], attention was called to the fact that the brightest variables have the longer periods, but at that time it was felt that the number was too small to warrant the drawing of general conclusions. The periods of 8 additional variables which have been determined since that time, however, conform to the same law."

Take note of Leavitt's last word in that sentence, which concluded the paragraph: *law*. By all accounts, Henrietta Swan Leavitt was a quiet, circumspect woman. But she knew what she had found. Would anyone notice this time around?

The Danish astronomer Ejnar Hertzsprung, then thirty-eight years old and working in Potsdam, Germany, immediately understood the implications of Leavitt's work. If the brightness of a Cepheid star was the product of the length of its cycle, then all stars with the same cycle must be equally bright, no matter how that brightness appeared to differ when viewed from Earth. By comparing absolute luminosity (the total radiant power a star emits in all directions) with its apparent luminosity (the radiant power received by a telescope on Earth), it would be possible to calculate the distance of that star from Earth. Until the mid-twentieth century, sailors on Earth depended upon lighthouses to calculate how far they were from land, based upon the fact that the beacons emitted a standard amount of light; Cepheid stars can thus be understood as celestial lighthouses. To be certain that Leavitt's findings were correct, Hertzsprung calculated the absolute brightness of thirteen other Cepheid stars in other parts of the Milky Way, and then compared the results with Leavitt's work. He then took the further step of calculating the distance from Earth of the Small Magellanic Cloud, coming up with a figure of 30,000 light-years. This was a greater distance from Earth than had ever been suggested for any astronomical body, but, as Christianson notes in his biography of Hubble, the figure was vastly shrunk in the article Hertzsprung wrote on the subject because of a typographical error that put the distance at 3,000 light-years. Nevertheless, the Danish astronomer was the first to use Cepheid stars to measure astronomical distances.

Henry Norris Russell, the head of the astronomy department at Princeton University, was also intrigued by Leavitt's paper, and without yet knowing what Hertzsprung had done, selected the same thirteen stars to study. Russell came up with very similar results. It would be up to Russell's most talented graduate student, Harlow Shapley, who had arrived at Princeton in the fall of 1911 after gaining a master's degree at Columbia, to take the next major step forward. Russell and Shapley had begun to work together on a number of problems, but in this case, Shapley went well beyond the initial work on Cepheids done by his mentor. He would spend the next several years searching for additional Cepheids. By 1918, he had plotted the period-luminosity relation of 230 such stars, many of them previously undiscovered. Leavitt's work, that of Hertzsprung and Russell, and Shapley's own calibrations showed across the board that Cepheids with the same cycle were identical. It was time to publish a paper setting forth a new view of a vastly larger universe.

Hertzsprung had estimated the Small Magellanic Cloud as being 30,000 light-years away, and Russell had suggested a figure of 80,000. Harlow Shapley went much further. According to his calculations, using Cepheid stars as standards, the Milky Way was 300,000 light-years across. And Earth was no longer near its center but on its outskirts. Many eminent astronomers were extremely dubious about this greatly expanded Milky Way, some actively aghast. Interestingly, though, Shapley's most vociferous opponent was not an old-school conservative astronomer, but a scientist with an even more daring interpretation of the universe. Heber D. Curtis of the Lick Observatory was the leader of a group of astronomers who had recently come to believe that the spiral nebulae, which remained gassy amorphous blobs even when viewed through the largest telescopes, were in fact other galaxies like the Milky Way. The Milky Way itself, Curtis calculated, was no more than 30,000 light-years across. Shapley's new Milky Way, by contrast, was so vast that it could plausibly include the spiral nebulae within its boundaries. Shapley himself had been leaning toward the multiple-galaxy concept as recently as the year before, but had now become convinced that the spiral nebulae consisted only of swirling gas rather than individual stars, and that the Milky Way encompassed the entire universe.

The disagreement between the two astronomers reached such a pitch that the National Academy of Sciences arranged a debate between them, to be held in April 1920 in Washington, D.C., on the subject "The Scale of the Universe." The debate was preceded by an awards ceremony, which meant that many of the major names in American science were present. At the head table there was a visiting guest from Europe named Albert Einstein, whose General Theory of Relativity had been confirmed the year before during an eclipse of the sun that allowed the "warping" of space to be observed. Henrietta Swan Leavitt was not there, of course. Despite the fact that much of the debate turned on her discovery of the period-luminosity relationship of Cepheid stars, and the measurements that discovery had made possible, she did not hold the kind of position or the degrees that would have entitled her to an invitation. Still, there were those sitting in the hall in Washington who regarded her as having an exceptionally brilliant mind, Harlow Shapley among them.

Another person absent from "the Great Debate" was thirty-year-old Edwin Powell Hubble, a young astronomer from Wheaton, Illinois, who believed that Curtis was correct about there being multiple galaxies. Hubble was at his post at Mount Wilson in California, where Shapley had also spent the last few years. The two men were colleagues, but not friends. Shapley was from Missouri and had a feisty farm-boy presence, while Hubble had been a Rhodes Scholar and came from an upper-class background. Shapley also disdained Hubble's credentials as an astronomer, seeing the slightly younger man's knowledge of the classics of literature, including many read in the original Greek and Latin, as the sign of a dilettante. It would be Edwin Hubble, however, who would take Leavitt's discovery of the period-luminosity relationship of Cepheid stars to its furthest extent.

In 1923, the new 100-inch (2.5 m) telescope at Mount Wilson was completed. For the first time, it was possible to see that the haze at the center of the Andromeda Nebula consisted of stars, not just gas. Harlow Shapley had been wrong. He had been wrong not only about the makeup of the nebulae, but about their distance from Earth. Building on Leavitt's discovery, and Shapley's own studies of more than 200 other Cepheid stars, Hubble calculated that the Cepheids he found in the Andromeda Nebula indicated that it was at least 900,000 light-years from Earth, and obviously not in any way a part of our own Milky Way, even taking into account Shapley's great expansion of its size. (Both technological and theoretical developments, including later work by Hubble, led to currently accepted calculations that Andromeda is more than 2 million light-years distant from us.)

Hubble, along with Einstein, came to be widely regarded as a scientist on a level with Galileo, Kepler, and Newton in terms of changing our view of the universe—all had created new paradigms. He did not get ahead of himself, however, as Harlow Shapley had done in 1920, even when his papers contained clear hints of the extraordinary directions in which he was moving. As astronomer Allan Sandage put it in an article written for the 1989 centenary of Hubble's birth, "It was Hubble's mastery of language that gave some of his papers such dominance over prior work by others." He had a gift for extending and consolidating the work of earlier astronomers, putting all the pieces of a puzzle together in ways that seemed irrefutable. Between 1910 and 1920, Vesto Melvin Slipher of the Lowell Observatory had shown that the spectral lines of nebulae were shifted slightly to either the blue or the red end of the spectrum, suggesting that the nebulae were either moving toward Earth (a blue shift, as in the case of the Andromeda Nebula) or away from it (a red shift, as in the case of the nebulae in the constellation Virgo). Building on this observation, Hubble was able to demonstrate in 1929 that except for a few nearby exceptions like Andromeda, all the galaxies in the universe were speeding away from one another, as though they were the result of an enormous explosion. Those that were moving toward one another, it subsequently became clear, were doing so because the gravity of one galaxy was sufficient to pull another toward it. Andromeda and our Milky Way will eventually collide and become a single galaxy—a cosmic event whose results elsewhere in the universe have been photographed by the Hubble Space Telescope, most notably in the case of the Cartwheel Galaxy in the constellation Sculptor.

Although the speculations by Heber Curtis and others that there were many galaxies beyond the Milky Way had caused great discussion in scientific circles for nearly two decades, it was Hubble who put the whole picture together in 1929. The cosmos consisted of untold numbers of galaxies, many as large or larger than our own Milky Way, "island universes" as Hubble called them, that were speeding apart at enormous rates. Hubble calculated that the clusters of galaxies in Virgo, for example, were moving away from Earth at 1,000 kilometers per second. That led to a new question: Why were the galaxies speeding away from one another? The eventual answer, which did not come into general acceptance until the 1980s (and is still questioned by some) was the Big Bang, the explosion of a primordial atom that created the entire universe.

When Einstein was developing his General Theory of Relativity in the mid-teens of the twentieth century, he found that relativity demanded that the universe be either expanding or contracting. No astronomer, however, would back him up on that, telling him that it was in fact static. He therefore developed a mathematical "fudge factor" called *the cosmological constant*. Once Hubble's paper was published, Einstein disowned the cosmological constant, saying it was the worst mistake he ever made. (Ironically, some quantum physicists in the 1990s suggested reviving Einstein's "mistake," since it would help to resolve complex problems that arose in respect to quantum physics.) Even as

Einstein was working on his 1916 paper on the Theory of General Relativity, of course, Henrietta Swan Leavitt's brief paper on the significance of the luminosity-period relationship of Cepheid stars had been in print for four years. At that period, there was little communication between the theoretical physicists of Europe and the astronomers in America. The key to the problem Einstein was facing in Germany had already been provided by a spinster lady with no degree in astronomy working away in a crowded office on Garden Street in Cambridge, Massachusetts. The key was not inserted into the correct lock by Harlow Shapley and others, but once Edwin Hubble was able to actually see the stars—especially the telltale Cepheid stars—in Andromeda, the door to a new concept of the universe lay directly in front of him. And he turned the key.

Sir Isaac Newton wrote in a 1676 letter to the English physicist and inventor Robert Hooke, "If I have seen further, it is by standing on the shoulders of Giants." That is true of all scientists. Sometimes even a mistake by a giant of science, like Shapley, can serve to further the explorations of another. Sometimes, too, a giant comes in disguise. It seems wholly unlikely that a nearly deaf spinster sifting through thousands of photographic plates, a woman who was officially merely a glorified clerk, should discover an essential key to an entire century of revolutionary science. But titles, position, and advanced degrees are not the only requisites for great accomplishment. A brilliant mind, dedication, and a willingness to do tedious work that supposedly more important people are too busy to undertake can lead to extraordinary breakthroughs. Leavitt's boss, Edward Pickering, had many honors in his lifetime, and the projects he instigated at the Harvard College Observatory are still bearing fruit. But it may be that his greatest success was to know a fine mind when he encountered it. The "computers" he hired, including Williamina Fleming, Annie Jump Cannon, and above all

Henrietta Swan Leavitt, do honor to him still. He did not, of course, give them as much credit as he should have, but in that period he was going against the grain to entrust women with such work in the first place.

Of his three star "computers," it is Leavitt who is best remembered. Almost every book written about the great advances in astronomy during the twentieth century takes note of her discovery of the special nature of Cepheid stars, and to this day that discovery is a tool of every professional astronomer. Leavitt died of cancer in 1921 at the age of 53. She had been in precarious health most of her life. In an obituary in the Harvard *Annals* published in 1922, Solon I. Bailey wrote, "Miss Leavitt inherited, in a somewhat chastened form, the stern virtues of her puritan ancestors. She took life seriously. Her sense of duty, justice and loyalty was strong. For light amusements she appeared to care little. She was a devoted member of her intimate family circle, unselfishly considerate in her friendships, steadfastly loyal to her principles, and deeply conscientious and sincere in her attachment to her religion and church. She had the happy faculty of appreciating all that was worthy and lovable in others, and was possessed of a nature so full of sunshine that, to her, all life became beautiful and full of meaning."

This is a charming and heartfelt obituary of a spinster lady. Although Mr. Bailey also noted Harriet Swan Leavitt's many years with the Harvard College Observatory, there is no suggestion that he realized he was writing about a scientist whose work would prove crucial to a new understanding of the entire universe. There is a crater on the moon that has been given her name, in tribute to her own great discovery and to all the deaf scientists throughout history—although if things were ordered differently, her name might more properly have been attached to a variable star in the Magellanic Clouds, a bright one with a very long cycle.

To Investigate Further

Christianson, Gale E. *Edwin Hubble: Mariner of the Nebulae*. New York: Farrar, Straus and Giroux, 1995. This is a splendid biography of a very great and very complicated man that gives the work of Harriet Swan Leavitt its proper due.

Weinberg, Steven. *The First Three Minutes: A Modern View of the Origin of the Universe*. New York: Basic Books, 1977. Although a great deal has happened in physics and astronomy since 1977 (including a 1979 Nobel Prize in Physics for its author), this remains a classic work on the developments that led to the Big Bang theory.

Lang, Harry G. *Silence of the Spheres: The Deaf Experience in the History of Science*. Westport, CT: Bergin and Garvey, 1994. This book, researched over many years, includes brief biographies of both Harriet Swan Leavitt and John Goodricke.

Note: There are a number of web sites that offer capsule biographies of Leavitt, Cannon, and Fleming. They contain some useful information, but unfortunately tend to be at odds with one another when it comes to both details and a full understanding of what these women accomplished. Be wary.

CHAPTER 4

Joseph Priestley

Discoverer of Oxygen

Have you had a carbonated drink today? Or used an eraser? Are you concerned about the effect of carbon dioxide on global warming? If so, the next time you pick up a glass, offer a toast to Joseph Priestley. He not only discovered how to infuse liquids with those delightful bubbles, but invented the eraser and called it rubber. He was the first to discover carbon dioxide and to document the process of photosynthesis. He demonstrated that graphite could conduct electricity, and explained how he had proved it in one of his numerous books, *The History and Present State of Electricity, With Original Experiments*, published in 1767. Although it would be another sixty-four years before Michael Faraday's invention of the dynamo would make possible the large-scale production of electricity (see chapter 5), the subject was an extremely popular one at the time, and Priestley's book was a great success that would go through five updated English editions and be translated into Dutch, French, and German. A friend of Benjamin Franklin (who got him interested in electricity in the first place), Priestley supported both the American and French revolutions, a stance that eventually forced him to flee England and settle in Pennsylvania. Another of his books, *History of the Corruptions of Christianity*, caused such a ruckus that it was publicly burned. Yet he was himself a clergyman, a leader of the dissident movement that would become Unitarianism. His theological influence was considerable, but it was as a scientist that Priestley made his most significant contributions. He was the first to describe the properties of ammonia, sulfur dioxide, hydrogen sulfide, and carbon monoxide. To top it all off, he is credited with the discovery of oxygen.

In our age of specialists, the range of Joseph Priestley's interests can seem almost preposterously broad. Some professional scientists and historians tend to look on him as a "dabbler," though few people in history have ever dabbled quite so successfully, and during his own time he was recognized as a major scientific figure. While he had taken courses in mathematics at school, he was entirely self-educated in the sciences. A quintessential amateur, he spent his life pursuing whatever interested him, and over the years a great many subjects triggered his curiosity. Whatever his interest might be at a particular time, however, he was very focused in researching it, and quick to publish his findings—which is why he is credited with the discovery of oxygen instead of three other men who might have laid equal claim to the discovery, and in two cases did just that.

Joseph Priestley was born near the city of Leeds in northern England on March 13, 1733. His family were strict Calvinists, although following the death of his mother in 1739 and his father's second marriage, Priestley was adopted by his aunt, Sarah Priestley Keighley, who was more open-minded. Calvinist doctrine, formulated by the French theologian John Calvin in the 1530s, was one of the strongest strains of early Protestantism. It would prove too inflexible for Priestley, who became the pastor to a dissident church in Leeds in 1767. The beliefs of such groups, which could be found throughout England, and subsequently America, led to the rise of the Unitarian Church, which not only rejected the Catholic doctrine of transubstantiation (in which the wine and the wafer offered during Communion became the blood and flesh of Christ), but went so far as to deny Christ's divinity. Priestley's adoption of Unitarian (then called Socinian) beliefs was in accord with his utilitarian approach to scientific theory and experimentation.

Unitarians tended toward very liberal political views from the start, so it was hardly surprising that Priestley should hit it off so well with Benjamin Franklin when the two met in London in 1766. Franklin had first spent a year in England when he was only eighteen, and in subsequent years had developed numerous friendships with influential members of British intellectual society. He had come to England this time to appeal, successfully as it turned out, for Parliament to repeal the Stamp Act so hated by the American colonies. Franklin was known not only in England but throughout Europe for his famous kite experiment demonstrating that lightning was a form of electricity. In fact, his initial published suggestion concerning this experiment was carried out in both England and France before he himself attempted it in 1752. That led to being awarded the Royal Society's Copley Medal in 1753. This was the same prestigious prize that would be given in 1783 to John Goodricke, whom Priestley taught at the dissident Warrington school (see chapter 3), and which Priestley himself would receive in 1773. Priestley was fascinated by Franklin's scientific experiments, and would become a lifelong friend of the American statesman and inventor. Franklin, in turn, encouraged Priestley to carry out his own experiments, first on electricity, and subsequently on gases, or "airs," as they were then called.

The interest in electricity that Franklin aroused in Priestley was strikingly rewarded within a year, when the clergyman was able to demonstrate that graphite conducted electricity. This was not a minor discovery—graphite is one of the crystalline forms of carbon, which is the chief element in electrical resistors to this day. In his *History of Electricity* Priestley paid many compliments to Franklin's own work on the subject a decade earlier, praising both the "simplicity and perspecuity" of Franklin's style as well as "the noble frankness with which he relates his mistakes, when they were corrected by subsequent experiments." Priestley himself would follow Franklin's example in terms of admitting mistakes. His successive volumes on his experiments with "airs" often took note of earlier lapses.

Priestley's shift from electricity to the subject of gases was largely a result of his wide-ranging curiosity. In 1767, he broke completely with the Calvinists and organized his own church in Leeds. In one of those odd serendipitous instances that crop up throughout the history of science, Priestley's new church was located next to a brewery, and he became intrigued with the nature of the "air" that hovered above the fermenting grain. He conducted an experiment that showed this gas to be different from normal air. Not only did it appear to sink to the ground around the vats, suggesting that it was denser than the air around it, but it also extinguished lighted wood chips that Priestley tested it on. As with all his experiments, Priestley wrote up the results, and thus gets credit for being the first person to identify carbon dioxide, although he did not give it that name. Nor was he able to explain the chemical processes at work. His experiments, however, provided the clues that would lead other chemists to start answering such questions in the years just ahead.

Priestley's approach to experimentation was straightforward. If he observed something new, he would take what he had learned and apply it to a different experiment. He injected carbon dioxide into water, and was delighted with the slightly tangy, bubbly mixture that resulted. Priestley quickly published a pamphlet on the subject, *Directions for Impregnating Water with Fixed Air, in Order to Communicate to it the Peculiar Spirit and Virtue of Pyrmont Water* . . . Pyrmont water was the natural sparkling water found in certain springs, the kind that now costs several dollars a bottle. This pamphlet, with its very explicit instructions, was popular with the public and so admired by scientists that it brought him the Copley Medal for 1773. He had in fact been a member of the Royal Society since 1766, an honor accorded him in connection with his work on electricity. At the end of the Preface to his pamphlet on creating sparkling water, Priestley wrote, "If this discovery (though it doth not deserve that name) be of any use to my countrymen, and to mankind at large, I shall have my reward. For this purpose I have made the communication as early as I conveniently could, since the latest improvements that I have made in the process; and I cannot help expressing my wishes, that all persons, who discover any thing that promises to be generally useful, would adopt the same method." (The original text of this document was printed using "f"s for "s"s, as was customary at the time.) There is a modesty to this statement that echoes that of Franklin in his scientific writings. It should be noted, however, that the habit of quick publication Priestley urged upon others served him in very good stead. Had it not been for this approach, he would have lost the credit for several discoveries, including, as we shall see, that of oxygen itself.

Up to this point, Priestley had supported himself by several means, serving as a clergyman, teaching at the Warrington school and elsewhere, and augmenting his income with his published works. These were not necessarily scientific works. While teaching at Warrington, he had written *The Rudiments of English Grammar* (1761) and *A Course of Lectures on the Theory of Language* (1762). Both of these works, as Robert E. Schofield notes in his entry on Priestley in *The Dictionary of Scientific Biography,* are highly regarded by modern language scholars for their insistence on correct usage. But he had also had some failures in terms of publications. His 1772 *History of Optics* did not sell well enough to compensate him for the money he had spent researching the book. Priestley had married Mary Wilkinson in 1762, and had a growing family to support. He was delighted, therefore, to be offered a sinecure in 1773, the year he won the Copley Medal, by

William Petty, the Earl of Shelburne. Priestley was well known among the British aristocracy—he had dedicated his pamphlet on carbonation to the Earl of Sandwich, First Lord Commissioner of the Admiralty, not merely to curry favor but also on the reasonable assumption that carbonated water could be of particular use to the Royal Navy.

The arrangement he now entered into with the Earl of Shelburne was ideal in terms of his current needs, and would in fact stand him in good stead for the rest of his life. It was quite common among the aristocracy to employ an artist, writer, or naturalist (as scientists were generally called) to serve as an intellectual companion. By this time Priestley was quite a catch, and he was able to command a very comfortable situation. Officially, he was the Earl's librarian and an advisor to the much younger family tutor. He was free, however, to spend a considerable amount of time on his experiments, with the bill being footed by Shelburne. The Earl was a controversial figure politically, and many historians suggest that he was interested in making use of Priestley's connections to both religious and political dissidents. He made no objection to Priestley's friendship with Benjamin Franklin or his support for the American colonies during the coming Revolutionary War, although Priestley failed to persuade Shelburne to openly support the American colonies. Priestley was paid a stipend of 250 pounds a year. He and his family were given a house near Bowood, the Shelburne estate in Wiltshire, and stayed at Shelburne House in London during the winter social season.

It was over the next several years that Priestley did his most important scientific work. In 1772, the Royal Society had published the first of Priestley's works on gases, "Observations on Different Kinds of Air." This paper, which has been described as "magisterial" by historians of science, covered early pneumatic experiments (*pneumatic* being the word then commonly used for all research on gases), including the isolation of nitric oxide and anhydrous hydrochloric acid gases (the gases released when water is removed from crystals). In the previous decade a number of investigators had become interested in "airs," and although Priestley clearly benefited from their work, he was the first to test a wide variety of substances, often in new ways, to see what kinds of gases they might release. Sometimes he heated substances, as others were doing, but often using a new "burning glass," which was essentially a magnifying glass that focused sunlight on a narrow field, a tedious process that nevertheless could create a flame. At other times he mixed different substances together to see what would happen. He examined the residues left by substances that had been altered in variety of ways, experimenting with materials that had calcified, had developed vegetative growth of some kind, or had been subjected to an electrical discharge. Priestley often found that the proper equipment to carry out his experiments did not exist, and improvised with great success using common household items ranging from laundry tubs to clay pipes. As time went on, he called upon friends to supply or even manufacture laboratory equipment for him. Josiah Wedgwood, the founder of the great porcelain company, could be counted on to provide everything from ceramic tubes to mortars.

In the first volume of what became known as *Experiments and Observations on Different Kinds of Air* (1774), Priestley included illustrations of a portion of the room he used as a laboratory, and of a number of the apparatuses he had created or adapted for his experiments. Priestley did the drawings himself, as he had for the *History of Electricity*—he was also the author of a book on drawing technique, *A Familiar Introduction to the Theory and Practice of Perspective* (1770). These illustrations are very detailed, down to a mouse kept in a jar with a perforated top. He notes that when he was not working with the mice, he kept them on a shelf above his kitchen fireplace, to avoid chilling them unduly. Some of his devices were wonders of the practical imagination. As I. Bernard Cohen relates in *Album of Science*, there is an apparatus in the laboratory fireplace that was used to expel gas from various solids, "which were put into a gun barrel placed within the fire. The open end of the gun barrel was luted [attached with lute, a clayey cement] to the stem of a tobacco pipe, which led into a trough of mercury suspended from the mantelpiece. In this trough there is an inverted cylinder, full of mercury, in which the gas is collected." There were several such contraptions, which can sound like Rube Goldberg devices when described but did the job very well. It is clear that Priestley, like his friend Benjamin Franklin, had to be something of an inventor as well as a chemist. The laboratories in an American high school are high tech when compared with the improvised workplaces of Priestley and other eighteenth-century scientists.

In 1775 and 1776, Priestley wrote two more works on his experiments, and the three volumes were eventually given the collective title *Experiments and Observations on Different Kinds of Air*. Priestley's methods were not systematized in the ways that modern chemical experiments are. It was more a matter of one discovery leading to another. For example, his discovery of marine acid air (anhydrous hydrochloric acid) inspired him to attempt to produce similar reactions with other substances, resulting in the discovery of sulfur dioxide and ammonia. During this period he also discovered both nitrous oxide and nitrogen dioxide. Nitrous oxide is popularly known as laughing gas, because of the peculiar effects it has on people. In the 1840s an American dentist named Horace Wells of Hartford, Connecticut, discovered accidentally that it could be used as a form of anesthesia—this writer had a dentist, a professor at Tufts Medical School, who was still using it in the late 1940s.

Some sources credit the discovery of nitrous oxide to Humphrey Davy rather than Priestley, but this is incorrect. Priestley's experiment is well documented. He heated ammonium nitrate in a container with iron filings. The gas that was produced was then passed through water to cleanse it of any toxicity. It was

Davy, however, who first called it laughing gas. A chemist in Bristol, England, he conducted further experiments, testing its effect on respiration, for example, and found that it did make people behave in silly, giggly ways. Visitors to Bristol often sought Davy out to have the gas administered to them as a recreational diversion. While Davy was happy to oblige, he was also a serious scientist, and speculated that it might have anesthetic uses. The confusion as to whether Priestley or Davy originally discovered the gas is a minor example of the problems that can arise in correctly establishing the provenance of a discovery in any scientific field. But the Priestley/Davy debate is of minor import compared with the tangled story of the discovery of oxygen.

———

In the winter of 1771–72, a German pharmacist named Carl Wilhelm Scheele discovered that manganese oxide, when heated to a red-hot level, produced a gas. Scheele also noted that the gas produced bright sparks when it came into contact with heated charcoal dust, leading him to call the gas "fire air." In trying to explain the properties of the gas, he made use of the then widely accepted phlogiston theory. This theory had originally been promulgated by Georg Ernst Stahl (1660–1734), who was the physician to the King of Prussia as well as an experimental chemist. He believed that chemistry was defined by the "mixtive union," called the mixt for short, which had two basic elements, water and earth. There were three kinds of earth, however: mineral earth, evident in the heaviness of minerals; sulfurous or phlogistic earth, which was both light and flammable; and mercurial or metallic earth, which accounted for the brightness as well as the malleability of metals. As Bernadette Bensaude-Vincent and Isabelle Stengers write in *A History of Chemistry*, in the translation by Deborah van Dam, this meant "that the corrosion of metal and the combustion of wood or charcoal relate to the same phenomenon. The metal 'burns' (slowly). Corrosion causes it to lose its light and volatile phlogiston, as it causes mineral coal to do the same." The earth substances could be changed either by the solvent power of water or the heat of fire.

According to Stahl's theory, phlogiston was an entity, one that allowed a change to be effected in a substance but was itself destroyed in the process. The concept is clearly encapsulated in the word Priestley would use when describing the experiments that led to his discovery of carbon dioxide. When he placed a candle in a sealed container, he found that the candle would burn out very quickly, and stated that the candle had *injured* the air. A mouse in a sealed container would soon collapse, having also injured the air. Priestley found, however, that the phlogiston (as he saw it) could be reconstituted, and the injured air healed, by placing a plant in a container. This observation was the starting point later scientists worked from in revealing the complete process of photosynthesis. While Priestley achieved only a preliminary understanding of what was happening, his work once again paved the way for others.

To return to Scheele's experiment, it is important to realize that although his "fire air" was liberated through the burning of manganese oxide, his notes make it clear that he realized the same effect could be produced by heating mercuric oxide, which was the method Priestley would use. Scheele's "fire air" was oxygen, but he did not pursue further experiments, and even more crucially, did not publish his findings until 1777. A few months before Priestley conducted his own experiments, another pharmacist, a Frenchman named Pierre Boyen, also burned mercuric oxide and noted a release of gas and a concomitant loss of mass in the mercuric oxide. Boyen is seldom mentioned, but his name appears in an extremely useful chronology of the oxygen saga prepared by Fred Senese of the University of Maryland (see "To Investigate Further"at the end of this chapter). Senese points out that because Boyen failed to investigate further, he didn't realize that the gas was not ordinary air. Boyen later tried to claim he had been the first to discover oxygen, but his case was the weakest of all.

Boyen's experiment was carried out in April 1774. Priestley would conduct his in August of that year, also using mercuric oxide. Unlike Boyen, he went further, burning a candle in the air he collected, and was surprised to find that it had "a remarkably vigorous flame." Priestley immediately recognized that this was not ordinary air, and made careful notes that were soon published. He also wrote to Benjamin Franklin about his experiment, and asked Franklin about a story he had previously told. Franklin replied, "When I passed through New Jersey in 1764, I heard it several times mentioned, that, by applying a lighted candle near the surface of some rivers, a sudden flame would catch and spread on the water, continuing to burn for near half a minute." Franklin then discussed what we now call "marsh gas" for several paragraphs, relating his own unsuccessful experiments, and concluding by saying, "The discoveries you have lately made, of the manner in which inflammable air is in some cases produced, may throw light on this experiment . . . "

A month after Priestley's discovery, French chemist Antoine Lavoisier conducted an experiment with mercuric oxide and observed that the result was metallic mercury, but made no mention of gas being produced in his notes. Senese's chronology reveals that in October, Lavoisier, who was very well known by that time, received a letter from Scheele about his "fire air," but never replied.

At this point, the use of mercury in chemical experiments requires further comment. Mercury is the only metal that is liquid at normal temperatures. It is extracted from the sulfide cinnabar by roasting, and is poisonous, but no one knew that until the twentieth century. Mercury had fascinated human beings from antiquity and was much used in early chemical experiments. It was also used by hatters to cure furs. One of the results of mercury poisoning is dementia, which became common among hatters—Lewis Carroll's Mad Hatter is one of the few Wonderland creations based on an unhappy reality, one whose cause had not yet been understood. Carl Scheele, who died at the age of 43, is believed to have shortened his life considerably because of the dangerous materials he used, and often tasted. Priestley often did taste tests as well, but he may have been more cautious—or simply luckier.

In October 1774, Joseph Priestley accompanied the Earl of Shelburne on a trip to the continent and spent considerable time in Paris, where he met many notable figures. He was invited to dine with Antoine Lavoisier, and in the course of the meal described his experiments with mercuric oxide, detailing how they had been carried out and discussing the remarkable gas that had been released. Lavoisier, Priestley would later recount, "expressed great surprise." On his return to England, Priestley continued his experiments with the gas released by heating mercuric oxide. Using sealed containers (essentially bell jars), he demonstrated that a mouse could live longer in such a container filled with the special gas than in one filled with ordinary air. He described these experiments in his 1775 volume, and called his discovery "dephlogisticated air"—air from which the phlogiston had been removed.

Antoine Lavoisier, who had been born in 1743, was a great scientist, and is often called the "Father of Modern Chemistry." But he was not above a bit of highway robbery. He took Priestley's dinner table story about the gas he had produced from mercuric oxide and set about conducting his own experiments. In April 1775, he announced to the Royal Academy of Sciences in Paris that he had isolated an "eminently breathable" component of air from mercuric oxide. The biography *Lavoisier: Chemist, Biologist, Economist* by Jean-Pierre Poirier contains a telling story by one of Lavoisier's French colleagues, Edmond Genet, who on a visit to England met with Priestley. Priestley repeated his experiments for Genet, and the French visitor sent a report of them to the Academy in Paris. "At that time, Lavoisier was pursuing the same subject, and I was surprised on my return to hear him read a memorial at one of the sittings of the Academy which was simply a repetition on different words of Priestley's experiments which I had reported. He laughed and said to me, 'My friend, you know that those who start the hare do not always catch it.'"

Priestley would chastise Lavoisier in print, saying that the French chemist had gotten his idea from him. In the end, historians of science would give the credit for the discovery of oxygen to Priestley, with a secondary nod to Scheele, who in fact never claimed primacy, recognizing that he had delayed too long in reporting his discovery. Unlike Priestley, Scheele apparently much preferred laboratory work to writing it up, although he is credited with numerous discoveries, including those of chlorine, glycerol, and several amino acids. So Priestley did catch the oxygen hare, but he would be upstaged by Lavoisier in a larger sense. Lavoisier subsequently gave the gas its lasting name, oxygen. More important, he developed the bases of modern chemistry, both conceptually and in terms of nomenclature, demonstrating the principles of the conservation of matter, and constructing a sound system of the known elements, on which Dimitri Mendeleyev would build the periodic table a century later.

In the process of reinventing the concept of chemistry, Lavoisier demolished the phlogistic theories of Stahl. The discovery of the compound nature of water in 1784 by Henry Cavendish (who also discovered hydrogen and was the first to measure the specific gravity of both that element and carbon dioxide) was a further blow to phlogistical theory. But Priestley refused to accept the "new chemistry" of oxidation put forward by Lavoisier, Cavendish, and others. He found some genuine flaws in the new chemistry, proving that hydrochloric acid did not contain any oxygen, but chemistry was off on a new course, and Priestley's unwillingness to go along with it has been regarded by most experts as a major weakness. Robert Schofield points out, however, that Priestley said he would be willing to give up on the phlogistic approach if he could be shown with certainty that it was wrong. Carl Scheele was not even willing to go that far.

While Lavoisier has been anointed the "Father of Modern Chemistry," Priestley's point of view still surfaces in odd contexts. In 1962, the physicist and historian of science Thomas S. Kuhn introduced a concept that would become a major factor in how science would subsequently be viewed. The concept was that of the paradigm shift, which occurs when an entirely new view sweeps away everything before it, as in the previous chapter with Hubble's proof that the universe consisted of innumerable galaxies that were speeding away from one another. Beyond the phrase *paradigm shift,* which has become so much a part of the language that it gets (wrongly) applied to changes in women's fashion, Kuhn had some other startling views. He claimed that many of the old ideas may not be wrong, but simply discarded as the scientific establishment rushed to adopt concepts they found more useful. As an example of ideas that might not be wrong, he cited the belief, still prevalent in Shakespeare's time, that the disposition of human beings was dependent on "humors" contained in the body, such as phlegm—and of course we do still speak of someone as being "phlegmatic." Another example he used was Stahl's theory that heat is caused by phlogiston. There is "no standard higher," he wrote, "than the assent of the relevant community." In other words, truth is not the tally-keeper—acceptance by the majority of those working in a particular field is. There exists, for example, a minority among physicists who do not accept that the universe was created in a Big Bang. These physicists remain vocal, despite the fact that whole cadres of Nobel Prize winners constantly beat up on them. In this context, Priestley's stubborn refusal to buy the "new chemistry" of Lavoisier takes on a different cast.

———

In 1779, the relationship between Priestley and the Earl of Shelburne began to cool, and he left Shelburne's employ in 1880 for reasons that remain ambiguous, although the Earl continued to send him an annuity agreed upon in 1773. Settling in Birmingham, Priestley became a preacher at the New Meeting House, an exceptionally liberal congregation. He was asked to join the Lunar Society, whose members included his old friend Josiah Wedgwood; James Watt, inventor of the steam engine; and Erasmus Darwin, an esteemed naturalist whose nephew Charles would introduce the world to evolution. Many of these men were wealthy and helped support Priestley's research, which by now was largely focused on defending phlogiston against Lavoisier's attacks.

Lavoisier had eclipsed him in fame as a scientist, however, and Priestley devoted increasing energy to theological and philosophical matters. In 1782 he published *History of the Corruptions of Christianity*, and in 1786 a rationalist view of the New Testament called *History of Early Opinions Concerning Jesus Christ*. Such works made him the leader of the Unitarian movement, but also brought him many enemies. The 1782 book was soon banned in England and publicly burned in Holland. The 1786 book caused an even greater scandal, and was considered so inflammatory by the Church of England that some theologians warned that the foundations of the British Empire were at risk.

As though his religious nonconformism wasn't causing him enough problems, he openly supported the French Revolution when it broke out in 1789. Among Priestley's new London circle was the American revolutionary Thomas Paine, whose *Common*

Sense had been a central document of the American Revolution. Once the war with England was won, however, and the writing of the Constitution was begun, Paine caused so much trouble that he was virtually driven out of America, and took up residence in London. Paine also supported the French Revolution, but in a pamphlet so incendiary that even Priestley took offense. Charged with treason, Paine fled to France, became a French citizen, and was elected to the revolutionary Assembly. But being one of those agitators who is never satisfied for long, he ended up being jailed in France, and barely escaped with his life. Priestley was a far more rational dissenter, and his status as a scientist had long protected him from any grave repercussions. Nevertheless, even Priestley had become too controversial for his own good. In 1791, a mob gathered and burned Priestley's Birmingham church as well as his home and laboratory. The family had been warned about the approaching mob in time, and escaped.

Priestley and his wife found a home in Hackney, near London. Although many of his scientific friends had become wary of his religious and political stances, he was taken up by a new group of luminaries, including William Blake and Samuel Taylor Coleridge, and saw much of his old friend Benjamin Franklin, who was representing the new United States in England. During the early 1790s, all three of the Priestley sons left for America, and their parents followed in 1794, a voyage that lasted eight weeks in those days. They were welcomed to America by Governor George Clinton of New York, and after a whirlwind two weeks meeting prominent social and intellectual New Yorkers, made their way to Philadelphia. Entreaties that they settle there permanently were rebuffed, despite the offer of a professorship of chemistry, and they joined their sons in Northumberland. Northumberland was only 150 miles west of Philadelphia but was then on the edge of the American frontier.

The Priestleys began to build a new house, but before it was finished Mary Priestley died, in the summer of 1796, leading her husband to write that although the house had promised to be "everything she wished, she is removed to another." Priestley did travel to Philadelphia once a year, and became friends with both John Adams and Thomas Jefferson. Unsurprisingly, he supported the more democratically minded Jefferson against Adams in the election of 1800, and turned out to have backed the right man. Jefferson and Priestley corresponded regularly during the President's first term, and there are indications that Jefferson, although formally an Episcopalian, was increasingly intrigued with the tenets of Unitarianism. On April 9, 1803, Jefferson penned a long letter to Priestley about the latter's recently published disquisition comparing Socrates and Jesus. He invited Priestley to come stay with him at the White House, but by this time Priestley was too frail to make such a journey. He died peacefully on February 6, 1804.

Most great amateur scientists are known to us chiefly because of that scientific work. But Priestley was different. His theological treatises had great impact, particularly in America, where the first Universalist church to carry that name was founded in Gloucester, Massachusetts, in 1799. Some treatments of Priestley's life are chiefly devoted to his theology and the liberal political beliefs associated with it. Others concentrate on his numerous scientific discoveries. Some religious historians claim that Priestley would have wanted to be remembered primarily as a theologian, but others are quite certain that his discovery of oxygen was the high point of his life, as demonstrated by his unwillingness to let Lavoisier "catch that hare." He was certainly a very influential theologian, but he was also a major scientific figure. In fact, Joseph Priestley packed more accomplishments into a lifetime than most of us can even dream about, to the extent that it is difficult to see him whole, so brightly do the individual parts shine.

Brief biographies of Priestley, encompassing a few paragraphs, sometimes overemphasize the fact that he refused to give up his belief in phlogiston, and can even give the impression that his discovery of oxygen was something of a fluke. Most scientists, though, are more than willing to give him full credit for his remarkable experiments on air. The contest between Priestley and Lavoisier still reverberates in ways that intrigue both scientists and science historians. Two of America's most distinguished chemists, Carl Djerassi and 1981 Nobel Prize winner Roald Hoffmann, wrote a play about the subject called *Oxygen* that has been given several high-profile productions. Its action alternates between 1777 and 2000. The main characters in the 1777 scenes are Priestley, Lavoisier, Scheele, and their wives. The central figures in the contemporary section are scientists charged with deciding whom to award an "honorary" Nobel Prize for the discovery of oxygen. Like Michael Frayn's prize-winning play about quantum physics and the atomic bomb, *Copenhagen,* the plot offers opportunities to delve into questions that go far beyond chemistry itself. Both the authors of *Oxygen,* by the way, are winners of the American Chemical Society's highest honor: the Priestley Medal.

The year 2000 also saw the unveiling of a plaque, jointly bestowed by the Royal Society of Chemistry and the American Chemical Society, at Bowood House, Calne, Wiltshire, to commemorate Joseph Priestley's discovery of oxygen on August 1, 1774. There have been scientists in the past who openly disparaged Priestley's contributions to chemistry, clearly irritated by the fact that this English clergyman should be credited with the discovery of oxygen rather than Lavoisier, who named it and explained its properties in a much more complete way. With the placement of the plaque at Bowood House, however, it can at last be said that the debate is over. The scientific establishment has spoken: Joseph Priestley, the English clergyman, discovered oxygen.

In another sense, however, no plaque can really do Joseph Priestley justice. Let's go back to 1770, when he was already famous for his carbonated water but still several years away from the discovery of oxygen. Many Europeans were fascinated by a gummy substance secreted by a South American tree. The new world was full of wonders: corn, which Northern Europeans disdained but Italians found uses for; tomatoes, potatoes, chocolate, tobacco; all of them strange, too strange for some, but wonderful to others. And then there was this gummy stuff from some peculiar tree. Joseph Priestley, of course, managed to acquire some. What could you do with it, he wondered? He rolled it up into a ball and rubbed it against a page on which he had written some words in India ink that he was not satisfied with. And the words disappeared: rubbed out. "Rubber," Priestley dubbed the gummy substance, and the eraser was born. The eraser. A side issue, almost a silly one, in terms of Priestley's great accomplishments. In some ways, however, that application of a strange gummy substance from one distant continent to another already cherished substance from still another distant continent is evidence of the kind of throwaway genius that marks the very greatest minds.

To Investigate Further

Schofield, Robert E. "Joseph Priestley," *Dictionary of Scientific Biography*, Vol XI, pp. 139–47. New York: Scribners, 1975. There has not been a full-length biography of Joseph Priestley since the 1960s, and the two from that period, by Thomas E. Thorpe (1960) and F. W. Gibbs (1965), are difficult to obtain, although they may be in some public libraries. Robert E. Schofield's entry in the multivolume *Dictionary of Scientific Biography*, available at any good library, is the best account of its kind. Schofield was the editor of a comprehensive volume of Priestley's scientific correspondence published in 1963.

Bensaude-Vincent, Bernadette, and Isabelle Stengers, translated from the French by Deborah van Dam. *A History of Chemistry*. Cambridge, MA: Harvard University Press, 1996. This is a very well-written account of the development of chemistry from ancient Greece to modern times that emphasizes the historical context. It is hardly light reading, but those with some background in chemistry will find it fascinating.

Cohen, I. Bernard. *Album of Science: From Leonardo to Lavoisier, 1450–1800.* New York: Scribners, 1980. This book explains the history of chemistry through hundreds of illustrations dating from the period covered. Cohen, who became something of a legend as a Harvard professor, has written numerous books, and the text is concise and clear. There are only a few references to Priestley, but it is the kind of book that can get a reader excited about science in general.

Djerassi, Carl, and Roald Hoffmann. *Oxygen.* New York: Wiley, 2001. Praised by both Nobel Prize winners and theater critics, this play is a fascinating excursion into both science and the politics of science.

Note: There are numerous web sites that deal with Joseph Priestley. Some are slightly tongue-in-cheek, some very serious. Not all are entirely reliable. However, many excellent links can be found at www.dmoz.org (the Open Directory Project), including solid biographical information as well as facsimiles of two of Priestley's publications. The Fred Senese chronology cited in this chapter is another matter. It is difficult if not impossible to access directly, but can be found by typing *Antoine Lavoisier + oxygen* into a search engine, and scrolling down through the top 30 web sites to "General Chemistry Online: FAQ: History of Chemistry."

CHAPTER 5

Michael Faraday

Electromagnetic Lawgiver

In his mid-teens, around 1806, Michael Faraday became interested in chemistry. The son of a London blacksmith who was often in ill health, Faraday had a slightly older sister and brother, as a well as much younger sister. The family was extremely poor, and Michael received only a basic grade school education. At the age of fourteen he was apprenticed to a bookseller and bookbinder named George Riebau, who had a shop off Baker Street, quite near where the Faradays lived above a coach house in Jacob's Well Mews. Information about Faraday's life from the ages of fourteen to eighteen is sketchy, but it is known that at some point he purchased a popular new book on the subject of chemistry, Mrs. Marcet's *Conversations in Chemistry*—no doubt getting a good price from his employer—and that he did a very interesting thing with it: He took the volume apart and, using his new bookbinding skills, rebound it with a blank page interleaved between each page of text so that he could take notes as he gradually taught himself the rudiments of the science.

That incident alone suggests many of the qualities that would eventually make him the foremost English scientist of his time. It reflects the intellectual curiosity that would vault a barely educated London youth into the highest ranks of scientific achievement. It is indicative of the diligence and plain hard work that made it possible for him to learn so much on his own. And, far from incidentally, it demonstrates the manual dexterity that would sustain him through hundreds of experiments for which he often designed and built his own equipment. Like Joseph Priestley (chapter 4), Faraday's genius for constructing the apparatuses necessary for the exploration of entirely new aspects of science would serve as the foundation for his intellectual achievements.

Michael Faraday was born on September 22, 1791, in Newington, a Surrey village that has long since been absorbed by the sprawl of greater London. Less than a year earlier, his father had moved the family south from the area in northern England then known as Westmorland, in the hope of finding a better life. Due to his father's poor health, there was a constant struggle simply to feed the family, but London would give young Michael opportunities he certainly would not have had in the north. Even before the age of fourteen, he had begun running errands for Riebau, chiefly delivering and retrieving newspapers. Many people could not afford to buy a daily newspaper and Riebau would rent them one for a small price; once read, it would be passed on to another customer. Faraday would later recall the frustrations of this duty: "Often on a Sunday morning I got up very early and took them round. Then I had to call for them again, and frequently, when I was told 'The paper is not done with; you must call back,' I would beg to be allowed to have it; for my next place might be a mile off, and then I should have to return over the same ground again, losing much time and being very unhappy if I was unable to get home to make myself neat, and to go with my parents to their place of worship."

The Faraday family belonged to a religious sect known as the Sandemanians, which had been formed in a break with Scottish Presbyterianism. There were only a few hundred members of the sect, and since its members were firmly against any kind of proselytizing, this schism did not result in a major religious movement in the way that Joseph Priestley and his fellow dissenters gave birth to Unitarianism. Michael Faraday would remain devoutly religious throughout his life, but he saw his Sandemanian beliefs and his scientific pursuits as belonging to two entirely separate realms. It should be noted, however, that the Sandemanians believed that theirs was the only true church and that by following its strict moral code they were assured of salvation. These beliefs allowed members of the sect to endure considerable hardship with equanimity—a valuable asset for an impoverished youth with dreams of becoming a scientist. In addition, Faraday saw his investigations into what he called "God's natural kingdom" as being as important as the knowledge of God's spiritual kingdom revealed in the Bible.

Once Faraday became a full-fledged apprentice bookbinder, he moved into rooms above Riebau's shop, which he shared with two other apprentices. The other two young men eventually went into the theater, and biographers of Faraday suggest that the three apprentices must have had some lively times together. But Faraday clearly spent a lot of time reading. He would later write about the importance of Mrs. Marcet's volume on chemistry to his education, and how much pleasure it gave him to eventually meet her. She was a considerable figure in the first half of the nineteenth century. Her maiden name was Jane Haldimand; at the age of thirty she married a Swiss physician, Alexander Marcet, who had studied at the University of Edinburgh. She wrote her book on chemistry, her first of many on diverse topics, over a three-year period in which her first two children were born. It was written in the form of dialogues; after her mother's death, when she was fourteen, she had taken it upon herself to quiz her younger sisters following their classes with the family tutor, and to explain any points they were confused about. An enormous success, the book went through sixteen editions in England, starting in 1805, each also published in America. Copies of the early editions now bring several hundred dollars on the rare book market. For an 1833 revised edition, she would include a description of one of Michael Faraday's by then very famous experiments.

As he was discovering chemistry from Mrs. Marcet in his teens, Faraday was also inspired by reading the *Encyclopedia Britannica*, particularly its entry on electricity. But with his usual modesty, he would later write, "Do not suppose that I was a very deep thinker, or was marked as a precocious person. I was a very lively, imaginative person, and could believe in the *Arabian Nights* as easily as in the *Encyclopedia*; but facts were important to me, and saved me. I could trust a fact, and always cross-examined an assertion." Indeed, his first experiments were small ones outlined in Mrs. Marcet's book, which he carried out in order to test the validity of what he was reading. Despite Faraday's humility, it is difficult to view such an approach to reading as anything other than precocious, especially for a teenager who had received so little formal education.

Young Faraday took another step toward enlightenment by attending a dozen lectures on natural philosophy (which is what science was then called) between February 1810 and September 1811. He had seen an advertisement for these lectures while walking along Fleet Street, the rowdy center of London journalism. They were held at a private house under the auspices of the recently formed City Philosophical Society. Each lecture cost a shilling, which his brother, Robert, now a blacksmith himself, provided—Faraday's father died at the age of forty-nine in October 1810, and Robert became the head of the family. These lectures added to Faraday's growing store of knowledge, and brought him the acquaintance of several young men who would remain lifelong friends. He began exchanging letters with one, Benjamin Abbott, in order to improve his writing skills, and took lessons every week from another, Edward Magrath, who would tutor him in grammar two hours a week for the next seven years. Faraday also started taking lessons in drawing and perspective, so that he could better illustrate the notes he was assiduously taking at the Philosophical Society lectures. As James Kendall notes in his biography of Faraday, "His teacher was a French artist, Masquerier, who had once painted Napoleon's portrait but was now a refugee in London, living in a room above Riebau's shop. In return for this instruction, Michael used to dust Masquerier's room and black his boots." Some forty years later Faraday would visit the aged artist in Brighton, causing a mutual friend named Crabb Robinson who was present during the visit to write in his diary: "When Faraday was young, poor, and altogether unknown, Masquerier was kind to him; and now that he is a great man he does not forget his old friend."

Faraday's youthful efforts at self-improvement were to pay large, and surprising, dividends. He bound the illustrated notes he had taken at the lecture series, and Riebau was so proud of his apprentice that he showed the book to several good customers. One of these was a young man named Dance, who borrowed the bound notes and showed them to his father. The elder Mr. Dance was as impressed as his son, and gave Faraday tickets to lectures being given by Humphrey Davy at the Royal Institution. Davy, who had moved far beyond his days of experimenting with laughing gas (see chapter 4), was now, at thirty-two, head of the Royal Institution, and the recipient of a knighthood just the day before he began his four lectures on metals. Once again, Faraday took detailed notes, illustrated them, and bound them. They would prove to be his ticket to the future.

Sir Humphrey Davy was born in Penzance, Cornwall (the site of the famous Gilbert and Sullivan operetta *The Pirates of Penzance),* in 1778. He had been apprenticed to a surgeon-apothecary, and then gained a position as superintendent of the Medical Pneumatic Institution of Bristol, where possible medical applications of gases were being studied. Aside from his rather frolicsome adventures with laughing gas, he embarked on some important electrochemical experiments. In 1800, the Italian physicist Alessandro Volta created the first crude electric battery, which passed a current from two metal pieces through a chemical solution. A mere six weeks later, the English chemists William Nicholson and Anthony Carlisle proved that not only did chemical reactions produce electricity, as Volta had demonstrated, but that an electric current could be used to instigate chemical reactions. They performed the first experiment in what came to be called electrolysis, decomposing water into hydrogen and oxygen and proving that oxygen must weigh sixteen times as much as hydrogen.

Davy began his own experiments with electrolysis in 1800, focusing on metals, but although he grasped the fact that the elements of chemical compounds must be held together by electricity, it would be seven years before his experiments bore fruit. He finally succeeded in separating potassium from molten potash, and then sodium from common salt. Building on the work of others, Davy used electrolysis to isolate magnesium, calcium, strontium, and barium. It was these discoveries that would lead to his knighthood in 1810. In addition to his experiments, he had developed a great reputation as a lecturer. He had begun lecturing at the Royal Institution in London in 1801. Founded with the specific purpose of acquainting the public with scientific developments through public lectures, the Royal Institution soon found that the extremely handsome and dashing Davy was a major draw, and he was named to head it in 1807.

As Faraday approached the age of twenty-one in 1812, his apprenticeship as a bookbinder was drawing to a close. The Davy lectures had greatly inspired him, and he was determined to become a scientist. But with very little formal education and no money, he knew his prospects were dubious. He wrote to Sir Joseph Banks, the head of the Royal Society, begging for any kind of position in a scientific laboratory, but received no reply. He then recopied the notes he had taken at Davy's lectures, gave them a fine binding, and sent them directly to Davy, imploring him to give him a position at the Royal Institution, no matter how menial. But Davy, who had recently married (breaking the hearts of half the aristocratic young women in London, it was said), was on his honeymoon in Scotland. Faraday therefore accepted a post as a bookbinder with a French refugee named de la Roche, to whom he had probably been recommended by the artist Masquerier.

On his return to London, Davy examined the bound notes from the young bookbinder and was impressed. Although he was now ensconced in the upper rungs of British society, he had come from a family that, while not nearly as poor as the Faradays, had suffered its own financial difficulties—after his father's death, his mother had taken in boarders to survive. Faraday had indicated in his letter that he not only wished to pursue a career in science, but also to help his widowed mother financially. Touched, as well as impressed by the notes themselves, Davy granted Faraday an interview in January 1813. Though he was taken with Faraday, he had no position to offer, and could only promise that he would keep the youth in mind. In a bit of extraordinary luck, a position did open at the Royal Institution, in February. William Payne, who had been Davy's assistant for ten years, got into an argument with a Mr. Newman, the Institution's instrument maker, and physically attacked him. Payne was discharged, and Davy sent a note around to Michael Faraday at his new lodgings above de la Roche's shop on Weymouth Street, offering him the position. He would be paid twenty-five shillings a week, and given two rooms at the top of the Royal Institution. Thus a self-taught amateur chemist entered upon a career that would eclipse that of Davy himself. It has often been said, in fact, that Michael Faraday was Sir Humphrey Davy's greatest discovery.

———

As he began his new apprenticeship, Faraday wrote to his friend Benjamin Abbott about his various duties, all of which were fairly routine. He had been put to work extracting sugar from beets, a process that would provide the majority of sugar to the

European continent in subsequent decades. For one of Davy's experiments, he was asked to prepare a compound of sulfur and carbon. And he had "a finger" in helping prepare a lecture for another Royal Institution speaker. Faraday was more excited about participating in one of Davy's sometimes recklessly dangerous experiments, this one involving nitrogen trichloride, from which Davy was attempting to disintegrate nitrogen. Four successive experiments resulted in explosions. During the first of these Faraday was holding a vial, and the "explosion was so rapid as to blow my hand open, tear off a part of one nail, and has made my fingers so sore that I cannot yet use them easily." Fortunately, he was wearing a mask with a glass visor, but the glass itself was cut by the force of the explosion. Things didn't go any better with subsequent tries and, Faraday notes with some understatement, "with this experiment he has for the present concluded." Two months later, Faraday wrote a long series of letters about the art of lecturing to Abbott. These have been judged extremely perspicacious by historians, but, as Kendall puts it in his biography, "he writes without the slightest premonition of his own future fame."

No sooner was Faraday getting used to the routine—and occasionally explosive nature—of his duties at the institute, than he was offered what seemed the exciting prospect of accompanying Davy on a two-to-three-year trip throughout Europe and into Asia as his "philosophical assistant." This was a peculiar moment for such travels, since the Napoleonic wars were still very much in progress, with England and France at one another's throats. Napoleon had given Davy an award in the past, however, and granted him and his party a special traveling dispensation. Lady Davy, her maid, and Sir Humphrey's valet would complete the group. Just before they were to embark in October 1813, the valet, who was a refugee, announced that he could not go, since his wife feared for his life. That meant that Faraday became valet as well as scientific assistant, creating a decidedly awkward situation.

Lady Davy, who had been a wealthy widow when Sir Humphrey married her, was a snob of the first order, and proceeded to treat Faraday as a servant rather than as an assistant in her husband's scientific work. She made the young scientist's life hell, calling upon him to carry out a great many menial duties for her, and almost driving him to return on his own to London to resume his life as a bookbinder. Fortunately, there were sufficient compensations to keep him at Davy's side. Not only did he meet, and sometimes work with, the many leading scientists Davy visited in France, Switzerland, and Italy, but he also had the chance to broaden his knowledge of the world. One highlight was to see, in Florence, the telescope with which Galileo had discovered Jupiter's moons.

In the course of their journeys, Napoleon was deposed and imprisoned on Elba, but when he escaped and returned to power, Davy apparently decided the situation was getting dangerous, and the travelers returned to London in April 1815, more than a year earlier than had been planned. During the European trip, Faraday had become a genuine partner in Davy's experiments, joining with French chemists in Paris, for example, in a successful effort to identify iodine. He was now rewarded with a new position at the Royal Institution, as Superintendent of Apparatus and Assistant in the Laboratory and Mineralogical Collection, a promotion that brought not only an increase in pay but also better rooms at the institute.

At this point, Davy, "seduced by the delights of society," as John and Mary Gribbin put it in their brief but solid *Faraday in 90 Minutes*, spent far less time at the institute, giving Faraday a chance to shine more on his own. At first his scientific papers gave little hint of the splendors to come, but his reputation was growing. From 1816 to 1819, he returned to the City Philosophical Society as expert rather than student, giving a series of sixteen lectures on organic chemistry. Each year the number of papers he published in the *Quarterly Journal of Science* increased, from six in 1817 to eleven in 1818 and nineteen in 1819. He

made his first important discovery in 1819, identifying the chlorine and carbon compounds, one of which, hexaclorothane,
would eventually be used in fire extinguishers.

Davy made another two-year excursion to Europe in 1818,
and Faraday's importance to the running of the Institution increased. When Davy returned to London in June 1820 he found
that the president of the Royal Society, Sir Joseph Banks, had
just died, and Davy was soon elected to succeed him, which
meant that he had less time for the Royal Institution. The other
chief candidate for the Royal Society presidency had been one of
the wealthy board members of the Royal Institution, William
Wollaston. Wollaston would play an important role in the next
stage in Faraday's life.

That same year, 1820, the Danish scientist Hans Christian
Oersted discovered that when a magnetic compass needle was
held over a wire carrying an electric current, the needle moved
sharply to form a right angle to the current. When the magnet
was held underneath the wire, the needle moved in the opposite
direction, but once again formed a right angle to the wire. This
demonstration of a transversal electromagnetic effect intrigued
many scientists, including André-Marie Ampère in France and
Davy in England. Davy managed to demonstrate that a copper
wire carrying a current would attract iron filings, but it was
Ampère who succeeded in developing a rule concerning the relationship between the direction of the current and the direction
in which the needle would be deflected—to the left, for example,
if the compass is placed in the flow of current moving from the
positive to the negative terminal. Back in England, Davy's rival
for the presidency of the Royal Society, Wollaston, showed up at
the Royal Institution and asked for the use of equipment to test
a theory he had devised. He believed that a wire carrying a current would twist on its axis in response to the pole of a magnet.
The experiment failed—not because it was wrongheaded, but
because the equipment used was not delicate enough. Faraday

was not present during this experiment, but he came into the laboratory when Davy and Wollaston were discussing why it hadn't worked. This was in April 1821, and Faraday did not pay much attention—he was about to get married.

Michael Faraday married Sarah Barnard, the daughter of a silversmith, on June 12, 1821. She was also a member of the Sandemanian sect. The closeness of the Barnard and Faraday families would increase five years later when Sarah's brother John married Michael's younger sister, Margaret. With Davy's help, Faraday had gotten permission to set up housekeeping with Sarah in his rooms at the Royal Institution, a housing imperative since he was still earning only 100 pounds a year.

A few months after his marriage, Faraday was asked to write a history of the phenomenon Oersted had discovered in 1820, and in his usual methodical way, going back to his initial reading of Mrs. Marcet's book on chemistry fifteen years earlier, he repeated all the experiments that he would be writing about, including the failed effort by Wollaston and Davy, which he could not make work either. He suspected, however, that there was a circular aspect to electromagnetism, and devised a series of new experiments to demonstrate it.

Following some preliminary successes, Faraday took a deep basin and filled it with mercury. Some wax on the bottom held a magnet upright in this pool of mercury, with its north pole just under the surface. One wire, with an attached cork that would allow it to float on the mercury, was suspended above the basin, with its top end held in a tiny inverted silver cup that also contained a very small amount of mercury. Faraday then hooked a second wire over the edge of the basin, and both wires were connected to a battery. The first wire immediately started to revolve in a circle around the magnet. When Faraday reversed the poles of the magnet, the wire revolved in the opposite direction. This contraption, however primitive, was the first electric motor. It would be another ten years before Faraday would create a true dynamo effect, but this experiment on September 4, 1821, nevertheless was one of the great breakthroughs in the history of science.

It was not hailed as such at the time. Indeed, it got him into trouble. No one fully understood the significance of the experiment, and when he published his account of it in October, some people assumed that he was stealing Wollaston's idea, although what he had achieved was completely different from what Wollaston had set out to demonstrate. Suddenly, Faraday's poor background and lack of formal education were taken note of in a new way. Who was this lower-class upstart to think he could steal an idea from one of England's most admired men? Once he realized that there was a whispering campaign against him, Faraday wrote to Wollaston begging him to set things straight. Wollaston couldn't see what all the fuss was about—he at least understood that Faraday had done something different, and he saw no need to make any kind of public statement. He did, however, whether in a deliberate attempt to help Faraday or simply because he was curious about what the young scientist might do next, attend several of Faraday's lectures. Word that he was doing so also spread, and the accusations of plagiarism died down.

Davy made some effort to back Faraday up on the Wollaston matter. But then a similar problem, in reverse, developed between Davy and Faraday, with Davy indeed taking undue credit for Faraday's work on the liquefaction of chlorine. Many historians speculate that Davy was in fact becoming jealous of the success that his assistant was achieving, especially at a point when his attention to his own research was starting to lag. Faraday had been working on the chlorine problem while Davy was out of town, using chlorine hydrate. When Davy returned and was informed of these experiments, he suggested heating the crystals in a closed tube. He did not, however, make any statement as to what he expected to result from such a procedure. When Faraday produced an oily substance that turned out to be liquid chlorine, another chemist, Dr. Paris, was present in the laboratory, stopping by to see what Faraday was doing before proceeding to have dinner with Davy. Dr. Paris later documented that he had noticed oil in a tube, and taken Faraday to task for using a dirty tube. Faraday filed off the sealed end. An explosion took place, and the oil vanished, astonishing both men. While Faraday prepared to repeat the experiment, Paris left for his dinner engagement, during which he told Davy what had happened in the laboratory. Davy, Paris writes, "appeared much surprised." The next morning Paris received a note from Faraday saying that the oil was liquid chlorine. As soon as Davy was informed of that result, he ordered Faraday to test a number of other substances. Faraday wrote a paper on the subject, which was then reviewed by Davy and altered to make it appear that he rather than Faraday had made the original discovery, and that Faraday was merely assisting him.

Faraday would ultimately have the last word on this subject, years later, with Dr. Paris backing him up. That was after Davy's death, however. In the meantime, Faraday managed to downgrade the achievement—no matter who was getting credit—in an 1824 paper that showed chlorine had already been liquefied in 1805–06 by a chemist named Northmore. These various difficulties seem to account for the fact that when Faraday was proposed for membership in the Royal Academy in 1823, Davy, its president, opposed his election to this eminent body. The Wollaston situation was being used against him once more, but that only served to make Wollaston understand the gravity of the situation, and with his approval Faraday published a new paper that settled the matter for good, making clear the differences between the two experiments. On January 8, 1824, Faraday was elected with only one vote against him on the secret ballot. Whether that negative vote came from Davy is not known. It may not have, judging by the fact that when Davy was forced to relinquish his post at the Royal Institution in 1825 because of ill health, he suggested that Faraday be given a new post, and on February 7 the board named him director of the laboratory, "under the superintendance of the Professor of Chemistry," William Brande, with whom Faraday had been working closely for years.

Faraday's new position left him little time for his own research. Thanks in large part to Davy, the Royal Institution's laboratory was the best equipped in England, and carried out a considerable amount of research for commercial companies, whose payments were vital to keeping the institute afloat financially. During the early 1820s, Faraday worked on the development of steel alloys. A silver alloy was subsequently used in the manufacture of hearth screens, but the commercial possibilities of a so-called "chromium alloy" were not recognized. As Kendall points out in his biography, it wasn't until 1931 that some old samples stored at the institute were newly analyzed, revealing that the "chromium alloy" was in fact what we know as stainless steel.

Beyond his duties overseeing the laboratory, including a major effort to create new kinds of glass for use in navigational instruments, Faraday devised another way to increase the income of the Institution. He started a series of lectures, called the Friday Evening Discourses, at which he or other experts would discuss the latest scientific developments in various fields for a paying audience. In our own time, the public has become so used to new scientific wonders that it tends to take them for granted, but in the England of the 1820s scientific developments held an extraordinary fascination. It was all people could do to keep up with the pace of the Industrial Revolution, and the Friday lectures became an institution. They are still held today. That success led to the Christmas Lectures, begun in 1826 and also still given each year. These were particularly directed at children. Their parents could attend but had to sit in the balcony. Faraday himself would give more than 100 lectures at these two events between 1825 and his retirement in 1862. Many of these lectures have been preserved, often taken down in shorthand as Faraday spoke. His own favorite series of Christmas Lectures, *The Chemical History of a Candle*, has become a classic of scientific literature. In respect to Faraday's fame as a lecturer, it should be remembered that he spent seven years as a young man taking a weekly tutorial in English grammar from his friend Edward Magrath. During Faraday's early years at the institute, Magrath would even come to his lectures and give notes afterward on how to improve his presentation.

In 1824, a member of the board of the Institution endowed a chemistry chair specifically to make it possible to pay Faraday a little more of what he was worth. That endowment increased his annual income from 100 to 200 pounds. His remuneration would not increase again until 1853, when it was raised to 300 pounds. In the late 1820s, Faraday did a considerable amount of what we would call consulting work, for both commercial firms and the government. But he found that it left him no time for his own research, and refused such work after 1830. As one of the most celebrated men of his time, he could have commanded much larger sums as a professor at other institutions, and was offered such positions in 1827 by the newly created University of London and in 1844 by the ancient and illustrious University of Edinburgh. He turned all such offers down. In refusing the 1827 offer, he wrote, "I think it a matter of duty and gratitude on my part to do whatever I can for the good of the Royal Institution in the present attempt to establish it firmly. The Institution has been a source of knowledge and pleasure to me for the last fourteen years, and although it does not pay me in salary for what I *now* strive to do for it, yet I possess the kind feelings and goodwill of its authorities and members, and all the privileges it can grant or I require; and, moreover, I remember the protection it has afforded me in the past years of my scientific life." It is difficult to imagine anyone writing such a letter in the twenty-first century, when money can trump almost any loyalty. Many scholars believe that if Faraday had in fact taken another position, the Royal Institution would have collapsed financially. His efforts were keeping it alive.

Despite all his other duties, Faraday did manage to make some remarkable discoveries on his own from time to time. In 1825 he isolated a substance that was originally called "bicarbonate of hydrogen." At the time it was nothing more than a curiosity, but we know it as benzene, used in the production of dyes, rubber, plastics, and many other industrial products. It is also a known carcinogen and has been regulated by the Environmental Protection Agency since 1977. In terms of its uses, benzene was the most important of Faraday's chemical discoveries, but its importance pales next to the work he carried out in 1831.

Faraday had become increasingly interested in the mysteries of induction, which we now define as an alteration in the physical properties of a body that is brought about by the field created by another. The French scientist François Arago had performed an experiment in 1824 that mystified scientists across Europe. He demonstrated that when a compass was suspended above a rotating copper disk, the needle of the compass would be deflected. Since copper, unlike iron, is not magnetic, there seemed to be no explanation for the needle's movement. Faraday lectured about Arago's experiment in 1827, describing his own failed attempts to solve the mystery with new experiments. Then, from 1828 to 1830, Faraday collaborated with Charles Wheatstone on a series of lectures featuring Wheatstone's work on sound and musical instruments. One demonstration illustrated how a metal plate that was set to vibrating could cause another plate some distance away to vibrate also. This, it became clear to Faraday, was a form of acoustic induction. Some unseen force was in operation. He connected Wheatstone's work and that of Arago, and in 1831 began a new series of experiments on electromagnetism that focused on devising some principles of induction.

Previous experiments by Faraday and others had shown that a coil of wire in the shape of a helix could be turned into a bar magnet with north and south poles when an electric current provided by a battery was run through it. An iron rod inserted into the coil would become magnetized when the current was switched on. Faraday took this experiment further, constructing an iron "induction" ring with two coils of wire opposite each other. The induction ring was fifteen centimeters in diameter, about the size of a modern bread-and-butter plate. (The famous statue of Faraday at the Royal Institution shows him holding this ring.) What Faraday hoped and expected to find was that if current was switched on in one coil, the iron ring would serve to create an electric current in the second coil as well. The second coil was connected to a galvanometer whose needle would show any evidence of current.

He was surprised and disappointed to discover that the galvanometer needle flickered briefly only when the electricity was first turned on in the other coil, or at the moment it was switched off. Electricity was being generated in the second coil, but only at the instant the magnetic force was first surging or fading away; there was no flow of current being generated in between those instants. With his induction ring, Faraday had in fact created the world's first transformer. That was not enough, however; he wanted to achieve a steady current. By September 1831, he had found that he could achieve a short pulse of electricity by moving a magnet in and out of the center of a circuit. He began to suspect that the magnet itself was crucial. In October he wound a coil around a hollow paper cylinder. The coil was connected to a galvanometer. There was no second coil attached to a battery. All he had to do to generate electricity was to move a bar magnet in and out of the hollow center of the coil—but still he was not getting a steady current.

The Royal Institution had a large compound steel horseshoe magnet, but it was on loan to his friend Mr. Christie, at nearby Woolwich. Typically, Faraday did not ask Christie to return the magnet to him but instead traveled to Woolwich on October 28. He had already conducted eighty-four experiments over the past two months in his determination to solve the problem. He would conduct another forty-four that single day. He found what he had sought by using a copper disk, of the kind employed by Arago in 1824, suspended between the two sides of the giant horseshoe magnet. "The axis and the edge of the disc," he wrote in his notes, "were connected with a galvanometer. The needle moved as the disc turned. Effects were very distinct *and constant.*"

Michael Faraday, on October 28, 1831, invented the dynamo, or electric generator. It would be another half-century before Thomas Edison would begin to light whole cities with large-scale generators constructed on a different model. But the principle on which the new electrified world would be based was established by Michael Faraday on that late October day in 1831.

Faraday's discovery was immediately recognized as the pinnacle of his career. Even the British prime minister, Robert Peel, came to the Royal Institution to see a demonstration. Just as U.S. President Rutherford B. Hayes would later view Alexander Graham Bell's telephone as nothing more than a toy, Peel asked what possible use Faraday's contraption could be put to. Faraday, according to legend, replied, "I know not, but I wager that one day your government will tax it."

It would take decades for that to happen, but Faraday followed up on his discovery of the principle of the dynamo with numerous other experiments. Scientists had been wondering for a century whether such phenomenon as electric eels and static electricity, built up by the rustle of silk as a woman moved across a room, were caused by the same force. Faraday demonstrated that all electricity, from fish, silk, and batteries, was indeed the same thing. In a paper published in 1834 he introduced numerous terms that are used to this day. He called the liquid through which a current is passed *electrolyte*, and the process of decomposing it *electrolysis*. The terminal connections he named *electrodes,* which were of two kinds, an *anode* being positively charged, and a *cathode* negatively charged. The products of an electrolytic decomposition were named *ions*, Greek for travelers. The language used to describe electromagnetic experiments had been in chaos; Faraday codified it so successfully that his terminology was immediately accepted throughout the scientific community.

The dynamo principle was, in Faraday's view, just one step along the path to explaining how the forces of magnetism and electricity traveled through space. He conceived the idea that what he called "lines of force" must link magnetic poles, and form a field between them. Although he used the expression "lines of force" in 1831 and developed the concept over the next year, he was uncharacteristically cautious about publishing a paper on the subject. The idea of an unseen force field was revolutionary, and Faraday may not only have feared that it would be scoffed at, damaging his reputation, but may also have realized that to communicate the concept fully would require a use of mathematics that he was in no way capable of managing. Although he is recognized as one of the greatest scientific minds of all time, and as the single most towering figure among experimental scientists, his lack of formal education hampered him in respect to mathematics. He never mastered anything more than basic arithmetic. The prose he used in his papers was so lucid that scientists could easily grasp the nature of his experiments, and that lucidity was also invaluable in communicating his own and other people's ideas to the general public in his lectures. But his mathematical weakness undoubtedly held him back in dealing with an idea like force fields.

Nevertheless, he made sure to establish the fact that he had been the first to conceive of force fields, and therefore wrote a note in 1832 that was sealed and placed in a safe at the Royal Institution, to be opened after his death. Part of the note read: "I am inclined to compare the diffusion of magnetic forces from a magnetic pole, to the vibrations upon the surface of disturbed water, or those of air in the phenomena of sound: i.e. I am inclined to think the vibratory theory will apply to these phenomena, as it does to sound, and most probably to light." This statement makes clear that he was already thinking along lines that would lead to quantum physics in the twentieth century. He would not return to these ideas, however, until 1844.

Faraday had always worked extremely hard, and at times that took its toll. He and Sarah were unable to have children, much to their regret, but they doted on their nieces and nephews, who often could be found visiting them at the Royal Institution, either attending lectures or enjoying family games, including various forms of pitch-and-toss, using everything from marbles to walnuts, at which Faraday was very adept. He also loved the circus—when he succeeded at making an electric wire rotate around a magnet in 1821, he celebrated by going with his brother-in-law George Barnard to watch the horses circle a ring performing tricks. Nevertheless, his work sometimes brought him to the edge of collapse. Margery Anne Reid, the daughter of Sarah's sister Elizabeth, lived with the Faradays for much of the 1830s, and said later that when her uncle was becoming too distraught, Sarah would insist upon taking him to Brighton for some time off. In 1839, however, he had a serious nervous breakdown from overwork. It took a lengthy convalescence in Switzerland to repair his health.

After he had recovered sufficiently to return full-time to work, he gave a Friday Evening Discourse on the subject of lines of force in January 1844. This lecture further developed the ideas he had set down in the sealed note a dozen years earlier. The theory of atoms had gained many adherents by that time, but he insisted that such particles were products of the web created by a field rather than independent entities. In so doing, he was skipping ahead to ideas that would become the concern of quantum field theory a century later. Neither that lecture nor another in which he elaborated on these ideas in 1846 caused much reaction. Because of his spellbinding eloquence as a lecturer, those who attended were not discomforted, but it is doubtful that they understood what he was talking about. Nor did the scientific community pay much attention, either to praise or condemn him, at the time. In the 1860s, however, James Clerk Maxwell built his wave theories of light on Faraday's speculations about electromagnetic vibrations. A great mathematician, he was able to provide the mathematical framework that Faraday could not.

But while it would require the work of others to make the most of some of his insights, Faraday's three laws of electromagnetic induction and his two laws of electrolysis have withstood every conceivable test. The Laws of Induction are: (1) a changing magnetic field induces an electromagnetic force in a conductor; (2) the electromagnetic force is proportional to the rate of change in the field; and (3) the direction of the induced electromagnetic force depends on the orientation of the field. His two Laws of Electrolysis are: (1) the amount of chemical change during electrolysis is proportional to the charge traveling through the liquid; (2) the amount of chemical change produced by a substance by a given amount of electricity is proportional to the electrochemical equivalent of that substance. The induction coils that make ignition possible in automobiles are constructed according to Faraday's Laws of Induction, while the chrome plating on car bumpers is applied by using the Laws of Electrolysis.

Toward the end of the nineteenth century, James Clerk Maxwell would write, in reference to *Faraday's Lines of Force*, "After nearly half a century of labour, we may say that, though the practical applications of Faraday's discovery have increased and are increasing in number and value every year, no exception to the statements of these laws has been discovered, no new law has been added to them, and Faraday's original statement remains to this day the only one which asserts no more than can be verified by experiment, and the only one by which the theory of the phenomena can be expressed which is exactly and numerically accurate, and at the same time, within the range of elementary methods of exposition."

Faraday was not by nature a theorist, but the most empirical of researchers. Yet his experiments led him inexorably to concepts that are still being explored in the twenty-first century. He might have gone even further, but his mental powers began to wane by the late 1840s. He continued to give lectures, splendid ones, until 1860, but he was unable to carry out new research on his own, and found it difficult to follow new work by other scientists. He was increasingly confused during the last seven years of his life, but they were at least comfortable enough. Over the years he had turned down a knighthood and twice refused the presidency of the Royal Society. His religious beliefs forbade such trappings of power. But he had been persuaded in 1835 to accept a government civil list pension of 300 pounds a year, in addition to his salary from the Royal Institution. Lord Melbourne, the new Whig prime minister, offered it in a way that Faraday found insulting, however, and it was a major newspaper story until Melbourne agreed to apologize, and Faraday was persuaded that the money was his due rather than an honor.

Faraday resigned from the Royal Institution in 1861. He had already been provided with a house at Hampton Court by Queen Victoria, where Sarah cared for him as he gradually sank into senility. He died at the age of seventy-five on August 25, 1867. If he was loathe to accept honors during his lifetime, they were heaped upon him in the course of the twentieth century. He is safely enshrined among the greatest scientists in history. The centenary of his discovery of the dynamo was the occasion of major celebrations in Great Britain in 1931. For the 150th anniversary of that event, his face temporarily replaced Shakespeare's on the ten-pound note.

But the blacksmith's son lacking a formal education might have been most pleased by a salute from another great experimenter. Thomas Alva Edison, who also had no more than a grade school education, often said that Faraday's papers, books, and lectures were his greatest inspiration and taught him more than everything else put together, because the ideas were stated so clearly and cleanly that even an uneducated person like himself could readily understand them. Once Edison found the key to using Faraday's dynamo principle to light the world, it soon became apparent that Faraday had also been correct in his famous reply to a skeptical Prime Minister Peel. Governments around the world have indeed found a way to tax Faraday's monumental discovery.

To Investigate Further

Kendall, James. *Michael Faraday*. New York: Roy Publishers, 1956. Kendall, a past president of the Royal Society of Edinburgh, is a splendid storyteller, and this may be the most engaging biography of Faraday. Kendall had previously written a biography of Davy, and is very good on the complex relationship between the two scientists. This book is out of print, but is likely to be found in many public library collections.

Thomas, John Meurig. *Michael Faraday and the Royal Institution: The Genius of Man and Place*. Bristol, England: Adam Hilger, 1991. Written by the director of the Royal Institution itself, this is a thorough and readable biography, with numerous illustrations.

Gribbin, John and Mary. *Faraday in 90 Minutes*. London: Constable, 1997. This is one of a series of short biographies of great scientists written by these experienced popular science authors.

Simpson, Thomas K., with Anne Farrell, illustrator. *Maxwell on the Electromagnetic Field: A Guided Study*. New Brunswick, NJ: Rutgers University Press, 1997. For readers interested in seeing how Maxwell developed Faraday's ideas, this compendium of Maxwell's major papers, with commentaries, notes, and diagrams, as well as an introduction that places his work in its historical context, will serve the purpose well.

Faraday, Michael. *The Forces of Matter*. Amherst, NY: Prometheus Books, 1993. This is a scientific classic, a series of six Christmas Lectures delivered in 1859 that sum up Faraday's remarkable career.

Note: A number of Faraday's lectures are available on the Internet. The six lectures that make up *The Forces of Matter* can be found at www.fordham.edu/halsall/mod/1859Faraday-forces.html, at the Modern History Sourcebook, and the six lectures comprising *The Chemical History of a Candle* are available at the same source but with the suffix /1860Faraday-candle.html. There are numerous brief Faraday biographies available online; one worth accessing for its illustrations alone can be found at www.geocities.com/bioelectrochemistry/faraday.html.

CHAPTER 6

Grote Reber

Father of Radio Astronomy

One of the most impressive human constructions on Earth is the Very Large Array, located fifty-two miles (84 km) west of Socorro, New Mexico, on U.S. Highway 60. It consists of twenty-seven radio telescope antennas arranged in a Y pattern, with nine antennas along each of its three arms. Each of these dish antennas is eighty-two feet (25 m) across, and weighs nearly 230 tons. Because the site was integral to the 1997 movie *Contact* starring Jodie Foster and advertisements featured the star seated on the ground with the huge antennas rising behind her, most Americans are familiar with the spectacular nature of this scientific site. Far fewer people realize that these enormous instruments are descendants of a radio telescope built in a backyard in Wheaton, Illinois, in the 1930s by a visionary amateur scientist named Grote Reber.

Reber was born in Chicago on December 22, 1911, but grew up in the suburb of Wheaton. His mother was a schoolteacher in this well-to-do community. In one of those path-crossings that seem significant even though they are essentially accidental, one of her students was an older boy named Edwin Hubble, who would be the first to establish, in 1929, that the universe was filled with innumerable galaxies flying apart at great speed (see chapter 3). Her son's scientific interests were aroused by the growing field of electronics, and he attended the Illinois Institute of Technology in Chicago, taking a B.S. degree in 1933. He then took a job with a Chicago radio manufacturer as an engineer.

Reber was an enthusiastic ham radio operator in his spare time, a fact that is crucial to the role he would play in the development of radio astronomy. In 1901, Gugliemo Marconi had achieved the first successful transatlantic communications by radio. Commercial and governmental uses for radio communication developed rapidly, and by 1912 the first federal regulations went into effect. As David Finley of the National Radio Astronomy Observatory points out in his article "Early Radio Astronomy: The Ham Radio Connection," the amateur radio operators known as hams, "whose interest in radio was personal and experimental, rather than commercial, got the short end of the stick. They were given the use of wavelengths of 200 meters or shorter," while the long wavelengths believed to be the only ones that could be used for long-range communication were reserved for commercial and government use. Enforcement was lax until the United States entered World War I in 1917, when all amateur use was foreclosed, and regulations were stringent after the war. The amateurs had no choice but to see if they could discover ways to make better use of the short waves remaining to them.

Numerous ham experiments between 1921 and 1924 demonstrated that intercontinental communication could be carried out using ionospheric refraction—essentially bouncing the waves off the charged particles of the ionosphere, a section of Earth's atmosphere that begins at thirty-seven miles (60 km) above Earth's surface and extends outward to the Van Allen Radiation Belts, which mark the atmosphere's outer limits. These successes by ham radio enthusiasts attracted the attention of commercial enterprises ("naturally," Finley dryly remarks). But there was a drawback to shortwave communication that spared the amateurs from having even this wavelength snatched away from them: shortwave communication is plagued by both noise and static.

One of the companies interested in seeing what could be done with shortwave communication on a commercial basis was AT&T. At their Bell Laboratories complex in New Jersey, a young engineer named Karl Jansky was assigned the task of identifying the sources of this interference as a first step toward their possible elimination. Jansky, born in 1905, was originally from Oklahoma, but took a physics degree at the University of Wisconsin and joined Bell Labs in 1928. Enterprising and imaginative, Jansky proved the right person for the job. In a field he constructed an antenna that looked like a series of goalposts linked by Xs. The antenna was mounted on a turntable that allowed it to trace the direction of any radio signal, and immediately acquired the nickname "Jansky's merry-go-round." It was designed to receive waves of twenty-one megahertz, a wavelength of approximately fourteen meters.

At this point there are two technical matters that need to be explored. First, Hz, or *hertz,* is a unit of frequency, equal to a cycle per second. It is named for Heinrich R. Hertz, the German physicist who confirmed the electromagnetic theories of James Clerk Maxwell in a series of experiments carried out from 1886 to 1889, establishing that radio waves travel at the speed of light and can be reflected, refracted, and polarized as light can be. A megahertz—MHz—is equal to 1,000,000 Hz.

A second technical issue is the determination of wavelengths. Light, according to quantum physics, can behave both as particles and waves. Here, let's forget about the particles and stick with waves. Just as in the ocean, light waves have crests with valleys in between called *troughs.* The distance between one crest and the next (or one trough and the next), measured in a straight line, is the wavelength. Each color we see has a different wavelength, or we would not be able to distinguish between them.

(Color-blind people cannot detect some of these differences, and are thus unable to tell red from green, for example.) No human being can see the longer wavelengths that lie beyond those we register as red, such as infared, although in this case we can feel the heat that is produced. Microwaves travel on a still longer wavelength, and it is beyond them that we encounter radio waves. Light waves are extremely short—only 16 millionths of an inch at the violet end of the spectrum, up to 27 millionths of an inch at the red end. Radio waves that carry communications signals on Earth range from 5 to 20 feet (1.5 m to 6 m) for FM radio stations to 600 to 1,800 or more feet (209 m to 549 m) for AM stations. The electromagnetic radio waves that travel from distant stars are much longer, and although they are invisible, they also form a spectrum.

It was this electromagnetic spectrum that Jansky's merry-go-round antenna was designed to investigate. Over the course of several months, he identified three types of static, two of which were caused by nearby and distant thunderstorms. The third type was a faint hiss whose origin was unknown, and Jansky spent more than a year trying to pin down its source. His first experiments led him to believe it was coming from the sun, but as the months passed the direction of the strongest point moved away from the sun. Jansky noted that the signal was repeated every 23 hours and 56 minutes. This loss of four minutes each day is characteristic of fixed stars, and he was able to determine that the source must be at the center of our own Milky Way galaxy, in the neighborhood of the constellation Sagittarius.

There was a degree of luck involved in this discovery. Jansky happened to be working at a time when sunspot activity was at its lowest ebb—otherwise the signal he traced would have been obscured by the sun's magnetic activity. Jansky's discovery was made in 1932, and he issued a report titled "Electrical Disturbances Apparently of Extraterrestrial Origin" that made the front page of the *New York Times* on May 5, 1933. Subsequent articles on the subject by Jansky himself were published in 1933 by

Nature and *Popular Astronomy*. He wanted to continue investigating this phenomenon, but Bell Labs was primarily interested in whether static would interfere with transatlantic radio transmissions, and since it did not, Jansky was assigned to a new project. While professional astronomers saluted Jansky for his discovery, they did not feel that it was of any great significance. But if the professionals viewed interstellar radio waves as a mere curiosity, twenty-two-year-old Grote Reber was fascinated by them.

Reber had already made a name for himself as a ham radio operator, having managed to make contact with other hams in sixty different countries on every continent. Few people had succeeded in "working," to use the ham operator's term, the entire globe. He would later write that he was feeling that "there were no more worlds to conquer." Recording radio waves from outer space was certainly a new world, and Reber was determined to be a part of it. He applied for a job at Bell Labs, hoping to work directly with Jansky. It was the height of the Great Depression, however, and Bell Labs was not hiring any new employees, nor did they have any interest in pursuing Jansky's work further. Although Reber also approached several astronomical observatories trying to interest them in the subject, nothing came of it. He would later write that Jansky himself had told him, "The electrical engineers were not interested because they didn't know any astronomy and couldn't find anything useful in the subject. The astronomers were not interested because they didn't know any electrical engineering and considered their present techniques adequate for studies of the universe." Reber added that his own experiences confirmed these reactions.

While Jansky gave a number of lectures on his original discovery, the lack of interest among professional astronomers discouraged him to the point that he made no further efforts to explore the possibilities of radio astronomy. Reber, however, was not so easily deterred. He began to design and build his own radio telescope. Like Jansky, he recognized that his antenna needed to be able to seek signals from different parts of the sky, but he saw no need for a revolving base. Instead, his structure could be tilted to either the north or south. Earth's own rotation, he decided, would take care of the west-to-east sweep of the sky. Operationally, the tilt was changed with the use of a differential gear salvaged from a Model T Ford truck. He also turned to a very different kind of design for the antenna itself. Janksy had noted that radio signals from deep space were weak. In order to focus them better, Reber decided to construct a parabolic antenna—a "dish"—with a sheet metal surface that would reflect radio waves back up to a receiver located twenty feet (6 m) above it. The receiver was designed to enhance the signals by a factor of several million. An electronic stylus recorded the radio waves on charts.

Reber's dish was thirty-one feet (9.5 m) wide. He didn't have much choice in the matter, as he pointed out in a lecture: "In building the supporting superstructure the longest lumber available at the hardware stores in Wheaton was twenty feet, dictating a maximum diameter of 31 feet for the dish." A wooden tower also had to be built to provide access to the receiver. Reber did most of the work building his radio telescope by himself, with an occasional helping hand, over several months in 1937. The materials cost about $2,000, no small amount for that time. The structure was, of course, the talk of Wheaton—and would have been anywhere else in the United States or around the world. No one had ever seen anything like it, and it would remain the largest and most sensitive radio telescope on the planet until after World War II. Even the parabolic disks of almost all modern radio telescopes are no more than 300 feet (91.4 m) across; if they were any larger, Earth's gravity would warp them, distorting reception.

An *Astronomy Today* article by Sancar James Fredsti, a research engineer at the California Institute of Technology Owens Valley Radio Observatory, gives a particularly clear picture of the differences and similarities between radio and optical telescopes.

The noise, or static, that streams through the universe has a variety of signal properties, "such as frequency, phase, amplitude and in some cases repetitive patterns." Both rapidly spinning pulsars and very distant quasars, which emit as much energy as one hundred galaxies and are believed to form around black holes at the center of galaxies, give out signals that identify them as separate "point" sources. A different kind of signal is emitted by "field" sources, such as the vast clouds of gas and dust that act as incubators for new stars. As information from any kind of source is received, it must be mathematically analyzed to separate what is useful and significant from the surrounding noise.

Both optical and radio telescopes are collecting electromagnetic energy—the same theories apply whether that energy takes the form of light as viewed by optical telescopes or radio waves collected by radio telescopes. There is a major difference, however, in terms of the wavelengths that are being studied. As Fredsti puts it, "Optical telescopes operate at very high frequencies and microscopic wavelengths, while their cousins the radio telescopes work at lower frequencies and longer wavelengths." That means that the beam width of a radio telescope is about 200,000 times that of an optical telescope, with a commensurate reduction in resolution. Thus radio telescopes have their limitations. On the other hand, they can detect objects that are impossible to view with an optical telescope and can reveal the true nature of objects in ways that optical telescopes, even the Hubble Space Telescope, cannot. Pulsars, for example, were discovered in 1967 by Jocelyn Bell Burnell and Antony Hewish, working at the Mullard Radio Astronomy Observatory in Cambridge, England. Although pulsars can be detected visually, their incredibly fast spin cannot. The search for black holes, which allow no light to escape and thus are invisible, could not have been carried out without radio telescopes.

While many of the factors described above had not yet been grasped in 1937 when Grote Reber built the first dish telescope, he was well aware that he needed to compensate for the weak signals he would be collecting. That was the reason for the paraboloid design, but it also carried over into other features of his homemade radio telescope. Surmising that the objects Jansky had detected must be very hot, Reber concluded that they would be more easily detected at much higher frequencies. Jansky had used a 20-MHz (15 m) wavelength; Reber started out using a 3,300-MHz (11 cm) wavelength that he believed would provide much greater detail. In fact, this approach did not work. He built a second receiver with a 900-MHz (30 cm) wavelength, but still had no luck. On his third try in 1938, at a 160-MHz (1.85 m) wavelength, he detected radio emissions from the Milky Way, confirming Jansky's original work in greater detail.

Reber faced other difficulties in his work. The sparks from automobile engines created too much interference during the day for successful results, even in Wheaton, so he found it necessary to work almost entirely at night. These automobile sparks created fuzzy spikes on his charts even at night, but the results were still clear enough to be included in the articles on his work that he began publishing in 1940. He presented other data as contour maps, showing clearly that the strongest (brightest) signals in the Milky Way emanated from its center, corresponding with the bulge found at the center of almost all galaxies in later years. It wasn't until the late 1990s that astronomers came to the conclusion that almost all galaxies have a vast black hole at their centers, which creates enormous stellar activity as their massive gravity pulls star systems toward them. Early clues to this phenomenon were present in Reber's data, including bright radio sources in Cygnus and Cassiopeia that his efforts revealed for the first time.

Reber's first paper on his radio telescope discoveries appeared in the *Proceedings of the Institute of Radio Engineers*, #28, in 1940. This was a technical article describing the details of the instrumental design of his radio telescope and the method he had used to reduce and record the data he was receiving. It should be noted that he did not refer to his instrument as a "radio telescope." That term was not introduced until after World War II. Reber simply called it an "antenna system." A second article, titled "Cosmic Static," was published in *Astrophysical Journal,* #91, later in the same year. Brief and in some ways tentative as this article is, it marks the beginning of radio astronomy as a separate discipline. "Several papers," Reber began, "have been published which indicate that electromagnetic disturbance in the frequency range 10–20 megacycles arrives approximately from the direction of the Milky Way. It has been shown that black-body radiation from interstellar dust particles is not the source of this energy."

In a footnote, Reber directed readers to the articles by Jansky that had originally inspired him, as well as to a paper by two other Bell Labs engineers on the same subject, published in 1937. His reference to black-body radiation was footnoted with a citation of an article by Fred Whipple and Jesse Greenstein (with whom Reber would later collaborate), which had appeared in the *Proceedings of the National Academy of Science,* also in 1937. This second citation makes clear that Reber fully grasped the potential significance of what he was doing. Black-body radiation would later prove to be a key element in establishing a theoretical proof for the Big Bang theory of how the universe began, after Arno Penzias and Robert Wilson of Bell Labs stumbled on the steady hiss of cosmic background radiation in 1965 while they were developing a receiver for the first communications satellite, *Telstar.* This is not to suggest that Reber was thinking about what would come to be called the Big Bang itself, which was a mere gleam of a concept in 1940—in fact, as we shall see, he would later write a paper skeptical of the theory—but rather that he recognized that the radio waves traveling through space could prove important in resolving a host of cosmic issues.

In his "Cosmic Static" paper, Reber included a photograph of his antenna and gave a brief description of the receiving system, but referred readers to his Institute of Radio Engineers article for detail. His power supply and auto-recording device were also pictured, followed by three pages of graphs. One of these presented a series of five chart readings from 1939, one from May, two from October, and two from November. These are elementary charts, but they do show clear peaks and valleys indicating wavelengths; dotted lines were added to illustrate the flat lines that would have resulted if no signals were being received. Later his charts and contour maps of areas of the sky would become more sophisticated. Even so, these first simple lines were enough to intrigue a number of astronomers. Reber had succeeded in capturing the interest of the professionals to a greater degree than Jansky had.

Reber's long nights spent scanning the sky continued as America entered World War II. In 1942, a second paper titled "Cosmic Static" was published in the *Proceedings of the Institute of Radio Engineers,* #30. Here he presented an equation using seven defined elements he had devised to measure the intensity of cosmic static. Amateur though Reber might be, he was well aware of the protocols used in scientific papers, and became increasingly adept at making use of them as the years went on. He knew that to be taken seriously he had to obey the rules of presentation used by such publications.

In 1944 the *Astrophysical Journal,* #100, published Reber's longest article yet, reporting on subsequent work, also titled "Cosmic Static." The article began with an abstract summarizing his findings: "Cosmic static is a disturbance in nature which manifests itself as electromagnetic energy in the radio spectrum arriving from the sky. The results of a survey at the frequency of 160 megacycles per second show the center of this disturbance to be in the constellation Sagittarius. Minor maxima [peaks] appear in Cygnus, Cassiopeia, Canis Major, and Puppis. The lowest minimum is in Perseus. Radiation of measurable intensity is found coming from the sun."

The eight-page report that followed discussed these findings, illustrated by charts and contour maps. While Reber did not draw any theoretical conclusions and was careful to note the limitations of his apparatus, he became the first person to confirm that the sun itself was emitting longwave radiation—he had improved his apparatus sufficiently to be able to take some data during the day despite the electrical sparks of automobiles. He concluded by stating that cosmic static from the sun, while intense, could not be the sole source of the radio waves he was recording, or else a large area in Sagittarius would have to be visibly as bright as the sun, which was not, of course, the case. Jansky, it should be remembered, had at first thought that the source of the static he was trying to trace was the sun, but then discounted that possibility when he discovered that the source was instead close to the center of the Milky Way. J. S. Hey established that the sun did in fact emit radio waves in 1942. These waves were assumed to be thermal in nature, produced by the heat of the sun, in accordance with the theory of thermal radiation. Reber's studies, however, made it clear that this was not the case. The theory predicted that radio emissions would increase at higher frequencies, and Reber's data flatly contradicted that idea. Some other process had to be involved.

This puzzle was solved in the 1950s by Russian physicist V. L. Ginzburg's theory of synchronotron radiation. Electrons and other particles moving close to the speed of light within magnetic fields, he postulated, would create cosmic rays. This theory not only explained the cosmic radio spectrum, but would eventually be put to work in the design of giant particle accelerators, such as CERN in Geneva, Switzerland, a hollow cement ring 16.7 miles (27 km) in circumference. At such facilities, high-energy particles are made to collide with other particles, with the results of the collision being analyzed to reveal the behavior of subatomic particles. Particle accelerators are at the other end of the spectrum, in terms of scientific apparatus, from Grote Reber's dish antenna in Wheaton, Illinois. But it was his backyard work that helped set in motion the need to find new theories to explain the source of the radio waves he recorded with his hand-made radio telescope.

Even back in 1944, Reber's new paper caused astronomers to take his work more seriously. The great Dutch astronomer Jan Oort (for whom the cloud of comets beyond Pluto is named, as discussed in chapter 2), was first off the mark in attempting to build on Reber's discoveries. Despite the German occupation of the Netherlands, copies of the *Astrophysical Journal* were acquired by Dutch astronomers. Reber's 1944 article suggested a possible solution to a problem that had long frustrated Oort. His optical telescope studies of the rotation and structure of our Milky Way galaxy had been impeded by the many clouds of interstellar dust that block visible light on the galactic plane, limiting visibility in the direction of the galactic center to a few thousand light-years. Reading Reber's work, Oort realized that radio waves would penetrate the dust, making it possible to "see" not only the galactic center but even the other side of the Milky Way. If the spectral lines of radio waves from the galactic center could be established, the Doppler effect, which shifts the frequency of such lines due to the relative motion between a wave source and an observer, could be used to measure the velocity of cosmic gas clouds. That would make it possible to better estimate distances to the gas clouds at the galactic center, facilitating the mapping of matter in the galaxy.

A student of Oort's, H. C. van de Hulst, was assigned the task of determining the possible frequencies of such radio spectral lines. Van de Hulst's mathematical model predicted that hydrogen should produce radiation that would appear at a radio frequency of 1,420-MHz (a 21-cm wavelength.) His results were published in 1945, and Reber himself started to construct a receiver that could detect this predicted line in 1947, but soon turned to other projects. Dutch radio engineers had no better luck, and successful detection of twenty-one-centimeter line radiation did not take place until 1951, when Harold Ewen and Edward Purcell of Harvard University overcame the problems involved. Hydrogen, both the simplest and most abundant element in the universe, forms the principal constituent of the sun and most stars, which coalesce from gas clouds. There is a great deal of hydrogen left over even after star formation in a given area, however, and throughout the universe its presence can be detected at the 1,420-MHz wavelength, as van de Hulst had predicted.

Oort's standing in the astronomical community was such that his interest gave Reber's work a new visibility and credibility. World War II had brought about many technical improvements in radio receivers, particularly in connection with the development of radar, that would prove to be of great value to radio astronomy. A new generation of electrical engineers with very open minds came to the fore, and the foundations of radio astronomy that had been laid down by Reber were quickly built upon. As new, larger radio telescopes were constructed around the world, it became increasingly clear that "radio eyes" complemented the "visual eyes" of optical telescopes: each could detect data that the other could not. The findings of radio telescopes began to be used as a guide to determining which galaxies optical astronomers should make the object of particular study, and as the data from both sources were compared and analyzed a far more complete picture of the way the universe worked started to emerge.

During the 1950s, radio astronomy faced two serious problems, however. The first had to do with physical limitations on the size of the antennas that could be built. It was obvious that the larger the antenna dish, the greater fineness of detail that could be recorded. However, because of the weight of the metal used to line the dish and the need to be able to move the dish to observe different parts of the sky, it soon became clear that no dish could be more than about 300 feet (91 m) in diameter. Indeed, the first dish of that size built by the National Radio Astronomy Observatory eventually collapsed of its own weight. The metal plates making up any dish could also warp over time, distorting the signals it received. One solution to these problems was the famous Cornell University dish, 1,000 feet (305 m) across, built at Arecibo, Puerto Rico, which substituted metal mesh for plates, cutting down on both the weight of the structure and the likelihood of warping. The ultimate solution, however, made possible by computers, was the Very Large Array (VLA) in New Mexico, originally conceived in the 1960s, constructed during the 1970s, and dedicated in 1980.

The twenty-seven antennas of the VLA can be moved on railroad tracks to any of seventy-two stations along the three arms of the system, with a flexibility that allows a separation between antennas as great as twenty-two miles (36 km) or as little as one mile (1.6 km). Every four months the antennas are moved to a new configuration, a process that takes about two weeks. Four standard configurations are used, which means that it takes sixteen months to complete the regular cycle. About 400 scientists a year, from institutions around the world, use the VLA; peer review is used to allot observing time, with a percentage reserved for graduate students engaged in thesis work, so that the facility is used for training as well as research by long-established astronomers.

The Very Large Array has a cousin, the Very Long Baseline Array (VLBA), which was dedicated in 1993. This use of identical radio telescopes at ten widely separated places, from St. Croix in the U.S. Virgin Islands across the continental United States to Mauna Kea, Hawaii, is based on *interferometry*, a technique originated by the Nobel Prize–winning American physicist Albert Michelson in 1887. An interferometer is a device that splits a beam of light into two parts that reveal an interference pattern of light and dark bands when they are recombined. That pattern allows the measurement of very precise details; laser interferometers are used, for example, to check the accuracy of machine tools. The VLBA uses interferometry on a much larger scale, with each of the ten separated radio telescopes being focused on the same galaxy or other celestial object. The signals are recorded on magnetic tape and then recombined by computer at the National Radio Astronomy Observatory facilities at Socorro, New Mexico. The result is equal to what could be obtained by a single dish with a gravitationally impossible diameter of twenty miles (32 km).

A second problem that existed in the 1950s has also been solved. Back then, enormous amounts of data were being recorded, but only on the same kind of paper charts used by Grote Reber. Literally miles of charts were accumulating that had to be examined section by section, a laborious and time-consuming process. In the 1960s, computers and magnetized film came to the rescue, making it possible not only to process data with far greater speed, but also to make comparisons between radio and optical data much more easily. Now the film of radio images could be superimposed over those of visual images to make the differences between them even clearer. The visual image of a spiral galaxy, for example, will show the light from the stars that compose it, while the radio image shows the gas clouds between the stars, aiding the determination of the density and size of the galaxy. Sometimes the radio emissions extend far out beyond the circumference of the visual spectrum, in vast jets of gas emanating from the center of the galaxy—a sign that there is likely to be a giant black hole at the center.

The development of radio astronomy in the sixty-odd years since Reber built his original dish has been extraordinary, but it also has its built-in problems. French astrophysicist Trinh Xuan Thuan writes in his book *The Secret Melody* about using the Very Large Array at Socorro; he always feels "an eery sense of unreality" when he does so. He notes that the radio telescopes are spread out in the desert over an area as large as the city of Paris and its suburbs, and that he is observing invisible objects with an instrument that he does not directly control. "I am entirely at the mercy of the computers," Thuan writes. "They point the 27 telescopes, combine the radio signals received by each telescope, digitize them, and manipulate them using complex mathematical operations, before displaying a color image on a television screen in front of me." It should be added that it is a "false color" image Thuan is looking at. The colors are added by computer to make clearer to the human eye such factors as the intensity of the signals. Thuan makes the point that even when computers are being used, the prejudices of the individuals who have programmed the computers can distort the results, and that for that reason any result must be confirmed independently by another person or team. He is not disparaging the amazing achievements of radio astronomy, but simply warning that scientists must be very careful not to let their machines "fool" them.

Like most amateur scientists, Grote Reber was by nature independent. As the field of radio astronomy took on great new importance from the 1950s onward, he continued to investigate the cosmos himself, but very much in his own way. In 1947, Reber took a new job with the National Bureau of Standards, and moved to the East Coast—that was one reason why he did not persist in the search for the twenty-one-centimeter hydrogen radiation wavelength. An application for a grant from the Research Corporation made in 1950 was awarded in 1951, and Reber resigned his job to work on a freelance basis and to build a new radio telescope on a mountaintop in Hawaii, on the island of Maui. The total grant was for only $15,000, but Reber managed to get the job done on that meager budget, hiring a few men to help him build the framework and haul it sixty miles (96 km) to its mountain site. From 1944 to 1950, Reber published another ten articles in various technical journals and popular magazines, including *Sky and Telescope* and *Scientific American*. These articles ranged from technical matters, as in "Antenna Focus Devices for Parabolic Mirrors," which appeared in the *Proceedings of the Institute of Radio Engineers*, to a general overview of radio astronomy for *Sky and Telescope*.

There was a hiatus in such articles between 1950 and 1954, when he began reporting on his findings from the new site in Hawaii. Over the rest of the decade several reports appeared in the *Journal of Geophysical Research*, as well as in the more widely read *Nature*. Reber wanted to explore very-long-wavelength signals that are generally blocked by the ionosphere. Only a few places in the world are free from such interference for any length of time. One of them is Tasmania, Australia's island state off the coast of Victoria. After three years in Hawaii, he relocated permanently to Tasmania, where his work continued to be backed by the Research Corporation. Although he continued his radio astronomy work, the experience of living in Australia led Reber to some new areas of inquiry, including studies of aboriginal culture, particularly in respect to their views of the heavens, and botanical experiments. An article about the latter, titled "Reversed Bean Vines," was published in the *Journal of Genetics* in 1964. His botanical work was conducted along the lines of Gregor Mendel's experiments on peas a century earlier.

During the 1960s, as radio astronomy emerged as an important field, Reber was honored with a number of distinguished awards. Ohio State University presented him with an honorary Sc.D. degree in 1962, a year in which Reber also delivered the Henry Norris Russell Lectureship of the American Astronomical Society. To top things off, 1962 also brought him the coveted Bruce Medal from the Astronomical Society of the Pacific. Endowed by Catherine Wolfe Bruce in 1898, the Bruce Medal is awarded each year for lifetime achievements in astronomy, with the honoree selected by a panel of six directors of observatories, three in the United States and three abroad. Among other winners over the years are such major figures as Hans Bethe, Arthur Eddington, Edwin Hubble, Edward Pickering, Allan Sandage, Harlow Shapley, Vesto Slipher, and Fred Whipple, all of whom are discussed in the course of this book. Nine years after Reber was awarded the Bruce Medal, it went to Jesse L. Greenstein, the founding head of the graduate program in astronomy at the California Institute of Technology, with whom Reber collaborated on a 1947 paper, "Radio-Frequency Investigations of Astronomical Interest." Reber, it should be noted, was only the second amateur to win the Bruce Medal, fifty-eight years after it went to William Huggins in 1904.

In the wake of these honors, Reber was asked to donate his original Wheaton radio telescope to the National Radio Astronomy Observatory. It was moved across country to the NRAO facility and museum in Green Bank, West Virginia, where it remains a major attraction, especially after being painted red, white, and blue for the nation's bicentenary in 1976. Reber himself oversaw its reassembly at Green Bank.

Reber's papers continued to appear over the next two decades, including one that caused a certain amount of fuss. Originally given as a lecture at the University of Tasmania in 1976, the paper was titled, "Endless, Boundless, Stable Universe." Reber, like a small band of major astrophysicists including Sir Fred Hoyle (another Bruce Medal winner) and Halton Arp of the Max Planck Institute, did not buy the theory that the universe was created in a Big Bang. The attack he launched on the theory in this paper is sometimes biting. The first words of the introduction are, "According to modern mysticism," by which he means Big Bang theory. Later he includes a graphic of a bumper sticker that reads THE BIG BANG IS AN EXPLODING MYTH and goes on to ask how such a myth got into textbooks. He explains, in quite technical terms, how his own research on long wavelengths creates problems for the Big Bang theory, and why he had been led to the conclusion that very different mechanisms were at work than assumed by Big Bang theorists.

Quite aside from his arguments against the Big Bang, this paper contains a telling paragraph about the position that any scientist, whether amateur or professional, can find himself or herself in when consulting experts in a given field. Reber is here describing the situation that arose when he discussed the possibility that longwave signals could be better recorded in Tasmania: "Asking for advice is a form of flattery: the recipient feels he must rise to the occasion. Advice is provided which under more sober circumstances would probably be declined. Also, most people have their own pet hobbies which envelop their lives. A stranger comes and proposes something different. Obviously it cannot be much good, or they would have thought of it first. Consequently the advice is negative."

Despite his heresy in respect to the Big Bang and his somewhat acid views of professional know-it-alls, Reber continued to be widely respected, not only for his historic contributions as the founder of radio astronomy, but because his grasp of technical problems of certain kinds was unparalleled. Thus he was one of a number of astronomers who were involved in experiments carried out by *Spacelab-2* in 1986, and contributed to several joint papers concerning the results. He was seventy-five then, and was still publishing papers in 1990.

Grote Reber's genius has been duly honored around the world. In addition to the spate of honors in 1962, he was awarded the Elliot Cresson gold medal of the Franklin Institute in 1963, the Jansky Prize of the NRAO in 1975, and the Jackson-Gwilt Medal of the Royal Astronomical Society in 1983. Great amateur scientists are not always fortunate enough to see the fruits of their labors achieve full maturity. Gregor Mendel's work (chapter 1), which would provide the foundation of genetics, was almost entirely forgotten at the time of his death. Harriet Swan Leavitt (chapter 3) knew that her work on Cepheid stars was becoming more and more significant before her death, but would not survive long enough to take delight in Edwin Hubble's eventual proof that there were untold numbers of island universes instead of just our own single Milky Way galaxy. Reber, on the other hand, saw his pioneering work develop into one of the greatest scientific tools in the history of humankind. Ironically, he came to disagree with one of the major conclusions that radio astronomy was used to support, the Big Bang theory. Yet the more scientists learn about the universe that Reber helped make it possible for them to see, the more new mysteries arise. It may yet prove that some of Reber's later work on longwave radiation will help to solve some of these problems in the future, perhaps even in ways that would confirm Reber's doubts about the Big Bang.

In pondering Reber's ultimate legacy, there is also the Search for Extraterrestrial Intelligence (SETI) to take into consideration. The idea of searching the universe with radio telescopes for signals from an alien civilization, first proposed by Giuseppe Coconi and Philip Morrison in 1959, has always been controversial. Right at the start, Coconi and Morrison noted that the probability of success in such an endeavor was difficult to estimate, "but if we never search, the chance of success is zero." While the development of radio astronomy made it clear that it would be possible for an alien civilization to broadcast its existence across the galaxy, most scientists scoffed at the idea, on several grounds. Some felt it was a waste of money and a frivolous use of the precious time available at radio telescope facilities, simply on the grounds that any such search was far worse than looking for a needle in a haystack. Our own galaxy has as many as 400 billion stars, and there are billions of radio frequencies.

There was also doubt among many scientists that any other planet in the galaxy, and perhaps the universe, could have developed life. The odds against life occurring elsewhere were calculated to be greater than the number of possible planets in our own galaxy, on the order of ten to the eighteenth power.

This argument can smack of self-importance, of course—"we're *unique!*"—but it was widely advanced until actual planets began to be detected around other stars in the late 1990s, and even then the possibility that the conditions on these planets allowed for the development of life seemed remote. SETI enthusiasts have always pointed out, however, that intelligent life of a very different kind from our own might exist, and that our definitions of life might be too narrow and self-serving. Even we humans are made up largely of electricity and empty space, and while all life on Earth is carbon-based, silicon is an even better conductor of electricity than carbon. During the last two decades of the twentieth century it was discovered that many varieties of life existed in places where it had long been believed impossible, in deep waters devoid of light and at temperatures that were considered too cold or too hot for any previously known form of life. That meant that a far greater range of planetary bodies might harbor life.

The first scientist to actually attempt a search for extraterrestrial life using a radio telescope was Frank Drake, using the eighty-five-foot (25-m) dish of the NRAO at Green Bank, West Virginia. He called his 1960 effort Project Ozma, after the princess in the Oz books, deflecting criticism with whimsy. Drake used the hydrogen frequency (1,420 MHz) discussed earlier—as the most common frequency in the universe, it seemed likely that any other intelligent life form would take advantage of that fact if it wished to broadcast a signal. Drake's project was limited in time and scope, examining just two relatively nearby stars, Epsilon Eridani and Tau Ceti, for a total of about 400 hours, with negative results, but the same so-called "magic frequency" has since been used many times.

Gradually, more astronomers came around to the idea that the search for extraterrestrial life was a respectable and worthwhile endeavor. In 1974 Drake and Carl Sagan, the astronomer and best-selling author, devised a message to be sent out into space from the Arecibo telescope, beamed in the direction of the giant star cluster known as M13. Using the digital binary system, the message contained information about the DNA that forms our genetic code, a diagram of our solar system, the population of the planet, and crude representations of the human shape and the Arecibo radio telescope itself. Carl Sagan also attempted to shift public opinion on the subject with his 1977 book *The Cosmic Connection* and his later novel *Contact*.

The central character in *Contact*, played by Jodie Foster in the film, was based on the radio astronomer Jill C. Tarter, who now heads the privately funded search by Project Phoenix. Such work was supposed to proceed under the auspices of NASA, but in 1993 Congress refused to provide a measly $2 million in funding. The work continues, however, and Drake, Tarter, and even some eminent scientists not officially involved with the search believe it will ultimately prove successful. As William J. Broad wrote in a 1998 article in the *New York Times,* "At the controls of the Arecibo search, Dr. Tarter became quite animated as she described futuristic arrays of hundreds and perhaps even thousands of small dish antennas tied to one another electronically, scanning the skies for the advanced civilizations she knows are out there."

If and when the day comes that we do manage to tune in a radio message from deep space, Grote Reber's backyard radio telescope in Wheaton, Illinois, will take on a significance that dwarfs all other achievements in the new field of science it created. Back in 1937, the longest lumber Reber could find in Wheaton was a mere twenty feet in length. It remains possible that those boards will prove to have formed the first step on the stairway to the discovery that we are not, after all, alone in the universe.

To Investigate Further

Sullivan, Woodruff Turner, ed. *Classics of Radio Astronomy*. Cambridge, England: Cambridge University Press, 1982. This collection of articles and scientific papers contains reprints of Grote Reber's several papers titled "Cosmic Static." A second book from the same publisher, *Early Years of Radio Astronomy,* was published in 1984. Both books will be of interest to readers who like to delve into original scientific documents. They can be found in many public libraries and are available from such used book sources as www.alibris.com.

Malphrus, Benjamin. *The History of Radio Astronomy and the National Radio Astronomy Observatory: Evolution Toward Big Science.* Melbourne, FL: Krieger, 1996. While it provides a linking narrative and is well-illustrated, this book focuses on technical explanations of the construction of radio telescopes and the discoveries that have been made with them. It is accessible to the general reader.

Sagan, Carl. *The Cosmic Connection: An Extraterrestrial Perspective.* Garden City, NY: Anchor Press/Doubleday, 1973. This is one of Sagan's most passionately argued books, in which he makes the case for the search for extraterrestrial intelligence.

Gutsch, William A. *The Search for Extraterrestrial Life.* New York: Crown, 1991. Written by the then chairman of New York's famous Hayden Planetarium, this book is written for older children, but many adults will find it a particularly clear explanation of the development of radio astronomy.

Note: There is a wealth of material on the Internet about radio astronomy, particularly at www.nrao.edu, including the David Finley piece quoted here. The Sancar James Fredsti article is available at www.astronomytoday.com. The first two "Cosmic Static" articles by Reber can be found at http://adsbit.harvard.edu.

CHAPTER 7

Arthur C. Clarke

Communications Satellite Visionary

Starting in 1927, at the age of nine, Arthur Charles Clarke set out on his bicycle every morning for school, pedaling nearly six miles from Ballifants Farm, Bishops Lydeard, to the town of Taunton in southwestern England. Taunton was an inland town, but it lay at the peak of a triangle between two famous towns on the English Channel, the port of Plymouth and the resort town of Bournemouth. All three of these place names had reappeared across the Atlantic Ocean in Massachusetts after the Pilgrims settled that faraway coast. Within a half-century after Clarke's birth in 1917, it became possible to flash a picture on a television screen from these towns on one side of the Atlantic to their namesakes on the other in an instant. The satellites that made this modern wonder of communications possible were the brainchild of Arthur C. Clarke himself.

By the time Clarke was thirteen, he was already spending some of his time dreaming about future wonders as he rode his bicycle across the English countryside, often in cold, wet weather. He had come across a science fiction magazine called *Amazing Stories* in 1928, and bought his first issue of *Astounding Stories of Super Science* in March 1930; that magazine had begun publishing only two months earlier. These magazines had brilliantly colored covers, but they were printed on the cheapest possible paper, and known as pulps for that reason. Pulp magazines came in many varieties, from romance to westerns, but it was the science fiction pulps that excited young Clarke. They were American magazines, and not always available. In *Greetings, Carbon-Based Bipeds!*, Clarke's compendium of his collected essays, he notes it was believed that these American pulps made their way to England as ballast in cargo ships, from which they found their way to Woolworth's, to be stacked in piles and sold for three pennies each. Clarke would spend his lunch hours pawing through the piles looking for copies of *Astounding*.

The Huish Grammar School in Taunton gave Clarke a sound basic education, including Latin, algebra, and geometry. His English teacher was a Welshman, Capt. E. B. Mitford, known as Mitty. Many years later, Clarke would dedicate one of his most famous collections of short stories, *The Nine Billion Names of God*, to this taskmaster. Mitty was the editor of the school magazine and was constantly trying to pry contributions to it from the students who, as Clarke has put it, "showed intimations of literacy." When Mitty demanded an article from Clarke, the teenager followed some other students to the Technical Institute in town, which he found to be full of strange machinery suggesting the laboratory of a mad scientist. One apparatus consisting of hand-cranked spinning disks made it possible to transmit the human voice over a beam of light, and that became the subject of Clarke's contribution to the Huish magazine.

Young Clarke added to his collection of *Astounding* whenever he could, but also read the classic works of Jules Verne and H. G. Wells. Some of these were available at the public library, but others he had to read at the local outlet of England's major bookseller, W. H. Smith. He could not afford to buy the books at Smith's, since they cost a shilling apiece, but like students the world over, he perfected the art of reading very fast while standing around in the bookstore during his lunch period.

Clarke recognized that Verne and Wells were a cut above some of the stories that appeared in the pulp magazines, but he was inspired by those stories, too, and has written about how advanced such stories could be in terms of the science they incorporated into their plots. He recalls a 1931 story by science fiction great Murray Leinster called "The Fifth Dimension Catapult," which dealt with quantum concepts of multiple dimensions. The idea that there might be more than the three dimensions we live in plus Albert Einstein's fourth dimension of time had first been suggested by the Polish mathematician Theodor Kaluza in 1919, but the concept was not seriously developed until the advent of string theory in the 1970s, and Leinster was way ahead of his time. Similarly, an *Astounding* editorial in 1933 discussed the theory of an expanding universe, an idea that was still very controversial at the time. Even the covers of the pulp magazines could be far ahead of current trends. In the 1980s, Clarke rediscovered that first issue of *Amazing Stories* he had seen in 1928. He recognized that the cover by Frank R. Paul, depicting the planet Jupiter as seen from one of its moons, had captured the huge swirling storm systems of the planet in ways that would not be seen until the Voyager probes of the 1980s, and that had not even been viewed through a telescope in 1928.

Clarke quite naturally began trying to write his own science fiction. In 1937 he started working on what would become the novel *Against the Fall of Night*, although he would not complete it until 1946. He also involved himself in some more practical aspects of science as a teenager, constructing homemade telescopes to get a better look at the moon. He used the successive Mecanno sets he purchased to construct numerous examples of futuristic engineering, regretting only that he never could amass the enormous amount of money, twenty pounds, to buy the most advanced #10 set. And by the end of the 1930s, Clarke had become an officer of the fledgling British Planetary Society, a group of science fiction fans and space travel enthusiasts.

In 1956, Richard van der Riet-Wooley stated in a speech for his investiture as Great Britain's Astronomer Royal that space travel was "utter bilge." When Yuri Gagarin became the first human to be launched into space in 1961, the British Planetary Society (by then renamed the British Interplanetary Society) put out a special bulletin announcing that Gagarin had been launched into "utter bilge." Arthur C. Clarke has commented that Riet-Wooley was given a bit of a hard time by Clarke's old pals at the Society, since most scientists were dubious about space travel in 1956. Clarke's own body of work simply assumed that space travel was inevitable, but Clarke has seldom engaged in put-downs, and has always been the first to point out his own mistaken assumptions.

Indeed, Clarke belittles his own ambitions following his graduation from Huish Grammar School. While he had made good grades, the family farm did not produce enough money to send him on to a university. A living had to be made, however, and he became a civil servant with His Majesty's Exchequer and Audit Department, comparable to the United States Department of the Treasury. He worked as a civil servant in London from 1936 to 1941. This left him with plenty of time to pursue his interest in science fiction. In his autobiographical book *The View from Serendip* (1977) he stated, "My life was dominated by the infant British Interplanetary Society, of which I was treasurer and general propagandist." His position as a civil servant made him exempt from military service, but in 1941 he decided to voluntarily enlist in the Royal Air Force (RAF), a move that he would later describe as "probably the single most decisive act of my entire life." He began teaching himself mathematical and electronic theory, and was made a radar instructor. There could have been no better place for him at this stage of his life. His talents suited the job, and he was of course enthusiastic about working in a new field of technology. On the other side of the coin, he would learn a great deal about both the theory and potential applications of advanced electronics. His response to what he was learning would lead to a short technical article, published in an obscure magazine, that would radically alter the future of human communications.

———

To understand the extraordinary atmosphere in which Clarke found himself working once he joined the RAF, it is necessary to go back to the mid-1930s, when the British made their first breakthroughs with radar technology. If it had not been for the development of radar, it is doubtful that England could have withstood the German bombing attacks during the Battle of Britain, and the country might have fallen to the Nazis before America entered the war in 1942. As it was, new radar technology during the war barely kept pace with counterefforts by the Germans. It was nip-and-tuck all the way, and Clarke found himself at the center of the most important realm of Allied technological invention aside from the work on the atomic bomb being carried out by the Manhattan Project in America.

England had been bombed during World War I, both by zeppelins (the hydrogen-filled airships that were the precursors of today's blimps) and by planes outfitted as bombers. Successive British governments in the 1920s and early 1930s fretted about the dangers of future bombing, but it was not until 1934 that any real action was taken in developing a defense. One of the first ideas put forward was to develop some kind of "death ray" emitted by a radio transmitter. Robert Alexander Watson-Watt (who was generally known simply as Watson-Watt), a Scottish engineer high in the Air Ministry, quickly recognized that while the prospects for a death ray were dubious, it might indeed be possible to produce radio technology that would alert the military that attacking planes were approaching the English coast. It had been noticed that radio waves produced an electronic echo when they bounced off an aircraft. If radio waves could be transmitted at a wavelength that produced such an echo at sufficient distance, an early warning would allow time for British fighter planes to get airborne quickly enough to try to shoot down the incoming planes. Watson-Watt's memo of February 12, 1935, "Detection of Aircraft by Radio Methods," establishes him as the father of radar, and began a period of feverish technical experimentation. The term radar, however, is of American coinage, an acronym for *RAdio Detection And Ranging*. The United States military adopted this term in 1940, but the British resisted it until 1943.

In his splendid 1996 book *The Invention That Changed the World*, Robert Buderi notes that Watson-Watt's famous memo took particular note of radio pulse technique. After emitting a radio wave, the transmitter would be turned off, so that the time it took to return to a receiver could be recorded. "Since radio waves move at the speed of light, the distance, or range, to the target then could be easily computed. For example, light travels at about 186,000 miles per second, meaning that it takes a radio wave pulse one microsecond to journey 0.186 mile, or 328 yards. An echo coming in ten microseconds after an emitted pulse would have traveled 3280 yards, or just under two miles." That meant that the incoming plane was one mile away, since the radio waves had to travel both outward and then back to the receiver.

Research was soon proceeding on two fronts. First there was the need to build a chain of radar towers on the ground along the Channel coast of Great Britain. The first four stations of what came to be called the Chain Home command were in place and put on twenty-four hour alert during the Munich crisis of September 1938, which concluded with Prime Minister Neville Chamberlain standing in front of 10 Downing Street on September 30, waving a useless agreement with Hitler and declaring, "Peace in our time." The Munich agreement allowed Germany to chop up Czechoslovakia, taking control of the most industrialized part of that nation, the Sudetenland. The out-of-favor Winston Churchill declared, "Britain and France had to choose between war and dishonor. They chose dishonor. They will have war." Even Chamberlain doubted that Hitler would keep his word, and the building of the Chain Home outposts was stepped up. By Good Friday 1939, the stations stretched from the Isle of Wight, the island in the English Channel off the south coast opposite Cherbourg, France, to the Firth of Tay in Scotland. The Munich agreement was abrogated by Germany on September 1, 1939, when fifty-three German divisions smashed into Poland. Great Britain declared war on Germany on September 3, and Churchill was brought into a new War Cabinet as First Lord of the Admiralty, a post he had held during World War I. A message was sent to all British ships at sea saying simply, "Winston is back." Eight months later, Churchill would be named Prime Minister.

Even before being brought into the government, Churchill had shown up for a demonstration of progress on the second radar front: creating devices small enough to be fitted into RAF planes. By the time war finally broke out, both land and air radar defenses were ready. The biggest breakthrough, however, would not come until February 1940, when the first test of a new invention, the cavity magnetron, was held. An electron tube that generates very high-frequency oscillations in the form of microwaves, the cavity magnetron achieved thirty times the output of previous radar devices. British electronics experts were already under tremendous strain as they worked on a host of wartime projects, and help was sought from America. The United States was given the top-secret details about the cavity magnetron in exchange for American technical help in further development of the device, particularly in respect to solving the difficult problems involved in making it usable in planes.

One of the Americans sent to England was Luis Alvarez, whose subsequent work during the war was enormously important. He conceived a blind-landing system for aircraft, the precision bombing radar known as Eagle, and a Microwave Early Warning radar set that greatly extended the distance at which enemy planes could be spotted. Alvarez would win the 1968 Nobel Prize in Physics for his development of the liquid-hydrogen bubble chamber, which makes the paths of charged particles visible. Years later, working in an entirely different field with his geologist son Walter, Alvarez put forward the idea that the dinosaurs were wiped out by an Earth-covering cloud of dust created by the impact of an asteroid or comet—a concept that was initially very controversial but would become widely accepted following the discovery of the enormous Chicxulub crater in the Yucatan Basin.

Arthur C. Clarke's 1949 novel, *Glide Path* (his only fiction that was not science fiction), centered on a character based on Alvarez, to whom the book was also dedicated; Clarke predicted in the novel that the Alvarez character would win the Nobel Prize by the late 1950s. Although it took another ten beyond that, Clarke had rightly recognized his genius during World War II. They met in the summer of 1943, when Alvarez spent six weeks at two Royal Air Force bases testing the Ground Controlled Approach system he had invented on every conceivable kind of RAF plane and using pilots of whatever rank. Clarke's own qualities had by then been fully recognized. When the project was fully mobilized following the six weeks of testing, he was the young officer entrusted with its operational control.

During his first two years in the RAF, Clarke had written a few short stories, but he would later note that he produced no writing at all during 1943—obviously his new responsibilities precluded that. Nevertheless, his exposure to the extraordinary minds involved in Great Britain's Radiation Laboratory ("Rad Lab") work provided him with the seeds for dozens of future stories and articles. As happened with many young men in both England and America who were too poor to go to college prior to World War II, the war itself in many cases offered opportunities for learning that no undergraduate university course of study could have matched. Few others of his generation made more of that situation than Clarke.

After the D Day invasion of Normandy in 1944, Clarke found the time to get back to his writing on occasion. As the European war drew to a close in the spring of 1945, he showed a draft of a scientific article he had written to a few friends, some of whom made suggestions to improve it. (The original copy is now in the National Air and Space Museum at the Smithsonian Institution in Washington, D.C.) This first version was titled "The Space Station: Its Radio Applications." It drew on ideas gleaned from the science fiction stories he had been reading since his early teens, as well as the futuristic concepts he had constantly discussed with his friends at the British Interplanetary Society, but it ventured into a new and more concrete realm as well. Making the most of what he had learned about electronics as an RAF officer who had been part of one of the great technological efforts of the century, Clarke gave an absolutely sound mathematical and scientific underpinning to an idea that could easily have been turned into a fictional story.

He made some revisions to his article, gave it the new title "The Future of World Communications," and sent it to an obscure technical publication called *Wireless World*. The editor of the magazine thought the title was a bit bland, and changed it to "Extra-Terrestrial Relays," coining the term "extraterrestrial" in the process. When the article appeared in October 1945, very few people aside from his personal friends took any notice of it. But Clarke himself, despite all the great science fiction stories and best-selling books he would write over the next half-century, has called it the single most important thing he ever wrote.

The subtitle of Clarke's article was a question: "Can Rocket Stations Give World-wide Radio Coverage?" Clarke proceeded to answer that question in the affirmative, suggesting that it would in fact require only three precisely positioned orbital relay stations to do the job. Before describing this proposal, Clarke used the first three paragraphs to discuss the limitations of Earth-based relay stations, noting that while ionospheric interference sometimes disrupted long-distance communication even in the case of telephone and telegraph service, matters were much worse in regard to televsion, for which ionospheric transmission was impossible.

Although television is thought of as a product of the 1950s, it had been discovered as early as 1908 that cathode tubes were capable of sending and receiving visual images, and they were used experimentally in Great Britain in 1934. The world's first television station, run by the British Broadcasting Corporation, began operations on November 2, 1936. In the United States, the pioneering work of Philo Taylor Farnsworth was trumped by a team organized by David Sarnoff, and television sets were available in the late 1930s, but the war brought a halt to the real commercial development of the new medium. It was not until the late 1940s that many performances were actually broadcast in either England or America. Clarke recognized, however, that television would be extremely important in the postwar world, and pointed out the vast expense that would be involved in building relay towers to convey the signal across any distance— at the time, 100 miles was the limit a signal could reach without being picked up and retransmitted. His article was an answer to that problem.

Clarke was at pains to make clear that the idea of launching rockets into an orbit around Earth was not as unreasonable as some of his readers might think. The German V-2 rockets, which had rained destruction on Great Britain, were the prototypes of more powerful ones to come, Clarke declared. It was by then known that the planned German A-10 rocket, with a transatlantic reach, would have attained half the velocity necessary (five miles per second) to launch it into an orbit around Earth. In a few years, Clarke suggested, it would be possible to build radio-controlled rockets that could "be steered into such orbits beyond the limits of the atmosphere." Later, manned rockets would be able to reach such an orbit, retaining enough fuel to break from the orbit and return to Earth. That would make possible the construction of a communications station in space.

To the general public, and even much of the scientific community in 1945, this vision of space stations was just science fiction, of a kind that had been around since Jules Verne's first novel, *From the Earth to the Moon*, published in 1865, and its 1869 successor, *Round the Moon*. Few people, however, understood in 1945 that Verne had correctly calculated the escape velocity necessary for a trip to the moon, as well as the ninety-seven hours required for the voyage through space. It would not be until NASA's Apollo program of the 1960s that readers would marvel at all the other things Verne got right: his three-man crew took off from the east coast of Florida and landed in the Pacific, a mere two and half miles from where the Apollo 9 capsule landed after its round trip to the moon in 1969. Clarke, as the former treasurer of the often ridiculed British Interplanetary Society, knew full well that most people would not take his proposal seriously. Therefore, to an extent that surpassed Verne's efforts, he set out to back his ideas up mathematically.

Clarke explained that a single orbiting station could provide coverage for half the globe. The entire Earth could be covered with only three such stations, one at 30 E longitude for Africa and Europe, one at 150 E for China and the oceans, and one at 90 W for the Americas. Clarke also noted that a greater number of orbiting satellites could be utilized for a greater variety of transmissions. He briefly described some of the technical considerations involved in the design of the orbiting transmitter, discussed the amount of power required, and speculated on the use of solar energy—so much more efficient beyond Earth's atmosphere—to provide power for the stations. In respect to this last idea, he carefully considered the loss of solar power that would take place at the times of the equinoxes when such satellites would enter into Earth's shadow, cutting off the sun's energy, and concluded that the total periods of darkness would not amount to more than two days per year, and less than one hour per day at any time.

In a summary, Clarke noted that (1) such orbiting transmitters were the only way to achieve worldwide coverage; (2) they would permit use of a bandwidth of one hundred thousand megacycles per second, and that the use of beams would make an almost unlimited number of channels possible; (3) power requirements would be very low, as would the cost of such power; and (4) the large initial expense would be less than for an Earth-based system, with much lower running costs over time. A two paragraph epilogue suggested that the imminent use of atomic-powered rockets, which he expected would be available within twenty years, made it clear that it would be foolish to spend money on Earth-based relays instead of waiting for the technology that would make orbiting transmitters possible. This epilogue is the only aspect of this article that caused Clarke to wince in later years. It should be recalled, however, that during President Eisenhower's first term (1953–1957), his administration was claiming that atomic power would soon make electricity so inexpensive that utility bills would amount to mere pennies a month. Clarke was far from alone in overestimating the possibilities of atomic power.

Clarke's concept of orbiting communications satellites was audacious and extremely well worked out, but what it makes it truly remarkable is that Clarke also suggested the best possible orbit for such satellites. He suggested an orbit of 26,000 miles (42,000 kilometers) above Earth, which would take exactly twenty-four hours to complete: "A body in such an orbit, if its plane coincided with that of the Earth's equator, would revolve with the Earth and thus be stationary about the same spot above the planet. It would remain fixed in the sky of a whole hemisphere and unlike all other heavenly bodies would neither rise nor set."

This is what is called a *geostationary* orbit. It is the orbit in which most communications satellites now reside and is usually referred to as a Clarke orbit, in honor of the visionary young man who suggested it. The exact altitude for such an orbit is slightly different than he worked out: 22,235 miles (35,784 kilometers). On the basis of what was known at the time, prior to the use of computers, his higher altitude was correctly figured, however. It should be noted that many articles and books refer to Clarke's orbit as *"geosynchronous* (geostationary)," as though the two terms were interchangeable. They are not. A geostationary orbit is one kind of geosynchonous orbit; it requires a location directly above the equator. There are synchronous orbits that are not stationary, although all must make one orbit per day, as Earth itself does, in order to be defined as synchronous.

Clarke has said that if he had patented the concept of orbiting communications satellites, he would be the richest man on Earth. At the time, however, he was equally interested in other aspects of his article. He emphasized the idea of the communications satellites as inhabited space stations that would be constructed in orbit. Although the American *Skylab* of 1972 and the Soviet *Mir* space station of 1986 partially fulfilled that vision, it was not until construction of the International Space Station began in 1999 that this aspect of Clarke's article began to attain full reality. In addition, Clarke added an appendix to his original article focusing on the rocket development that he believed would make his other concepts possible. As it would turn out, in the immediate postwar years, the only aspect of Clarke's article that would command the interest of governments was rocket development, and that was for military use. It would be seventeen years before the first test of a communications satellite would be attempted. In that time, however, Clarke would become a world-famous author.

———

Clarke was not dismayed that his article on satellite communication attracted little attention; too many things were happening on other fronts. He had already published a number of stories in British magazines, but finally cracked the American market when *Astounding Stories* published his novella "Rescue Party" in its September 1945 issue. Still, he was hardly earning enough money from his writing at this point to consider it a full-time career. His old civil service job was waiting for him, but he decided that he definitely did not want to spend the rest of his life as an auditor, despite the security such a position offered. He wanted to finally attend a university in order to fill the remaining gaps in his scientific education, and therefore resigned from the civil service before being mustered out of the RAF. But that step proved to be something of a catch-22. Because he had resigned from the civil service, he was denied a government scholarship. Fortunately, he had won an RAF essay contest with an article called "The Rocket and the Future of Warfare," which attracted the attention of a young member of Parliament, Capt. Raymond Blackburn. Blackburn pulled some strings and got Clarke a scholarship to King's College at the University of London, which he attended from 1946 to 1948, taking degrees in both physics and pure and applied mathematics.

Although he had written his communications satellite article while very much an amateur, he could now call himself a professional mathematician. During 1949 and part of 1950 he was employed as the assistant editor of *Physics Abstracts*, published by the Institute of Electrical Engineers, but it was beginning to look as though he could indeed make a living as a writer. He completed his first novel, *Prelude to Space*, during his 1947 summer vacation. He also became the chairman of the British Interplanetary Society, and in that capacity gave a speech in 1947 called "The Challenge of the Spaceship," which was widely reprinted and made its way into a collection of essays called *British*

Thought, 1947, bringing him a number of commissions from newspapers and magazines. Clarke even sent a copy of the essay to the grand old man of British literature, George Bernard Shaw, then ninety-one, who immediately sent back one of his famous pink postcards asking how he could join the Interplanetary Society. Shaw remained a member for the remaining three years of his life.

As the 1950s began, Clarke's writing career took off in a big way. In 1951, both the novel *Prelude to Space* and the nonfiction book *The Exploration of Space* attracted much attention, with the latter being selected by the Book-of-the-Month Club. The novel *Islands in the Sky* appeared in 1952, and 1953 brought his classic *Childhood's End*, which is still being purchased by thousands of new readers every year. Clarke has ruefully remarked that "an annoyingly large number of people still consider it my best novel." Now an established writer, publishing a book of fiction (sometimes a novel, sometimes a collection of short stories) as well as a nonfiction book almost every year, Clarke was in a position to do what he wanted and go where he wished. To the surprise of some people, this advocate of space travel now began exploring the world beneath the seas.

Introduced by Mike Wilson to skin diving in 1950, Clarke became increasingly fascinated by underwater exploration. From the mid-1950s on, with Wilson as his partner, Clarke spent more and more time diving in the Indian Ocean and writing books about his experiences. Wilson established a home in Colombo, Ceylon (now Sri Lanka), and Clarke spent nearly six months of the year with him and, later, Wilson's wife, Elizabeth. During the decade, despite a great deal of traveling, Clarke notes, "I appear to have written 140 pieces of fiction and 211 of nonfiction, a record never approached by me again." Many of the nonfiction pieces, Clarke says, were "ephemeral journalism." He wrote many reviews, not only of science and science fiction books, but also of the sudden spate of science fiction movies that began appearing in the 1950s. He was greatly impressed by the documentary-like George Pal movie of 1950, *Destination Moon*, but less happy with the producer's next effort, 1951's *When Worlds Collide*, because of technical lapses and a sappy Hollywood script. Despite his objections, George Pal subsequently showed him around the set of 1953's *War of the Worlds*. Ten years later, of course, Clarke would co-author with Stanley Kubrick the script of what most critics regard as the greatest of all science fiction films, *2001: A Space Odyssey*, based on Clarke's short story "The Sentinel," an effort for which the authors were rewarded with an Academy Award nomination.

As Clarke's career prospered during the 1950s, reality was beginning to catch up with his 1945 vision of satellites orbiting Earth. While in Ceylon in 1954, Clarke first heard the news, on shortwave radio, that the United States was preparing to launch a satellite during the International Geophysical Year of 1957–58. In October 1957, Clarke traveled to Barcelona as a member of the British delegation to the International Astronautical Congress. On the morning of October 4, he was wakened by a call from London's *Daily Express* seeking his comments on the launching of the first space satellite. The Americans had been beaten to the punch by the Soviet Union's *Sputnik 1*.

Sputnik 1 weighed only 183 pounds (83 kg). Designed to determine the density of the upper atmosphere, it had two radio transmitters, but their signals lapsed after twenty-one days, and the satellite itself, its orbit decaying, burned up in the atmosphere after two months. The second *Sputnik*, launched on November 3, was six times as large, weighing 1,113 pounds (504 kg), and carried the first live creature into space, a dog named Laika. Biological data sent back to Earth for a week showed that Laika was adapting to space, suggesting that manned missions could be safely attempted. The technology for bringing a satellite back to Earth did not yet exist, and Laika was put to sleep in space. *Sputnik 2* orbited the globe for 162 days before falling back into the atmosphere.

There was considerable shock in the United States and around the world that the Soviets has gotten into space first. This was not just a matter of scientific boasting rights—it meant that the Soviet Union was ahead of the United States in building rockets capable of launching a satellite, and that in turn suggested that they were ahead in terms of military rockets that could carry atomic bombs. The United States was further embarrassed when a Vanguard rocket developed by the U.S. Navy, carrying a test satellite, exploded two seconds after liftoff on December 7, 1957, the sixteenth anniversary of the Japanese attack on Pearl Harbor. Finally, on January 31, 1958, the United States successfully launched *Explorer 1*, which discovered the Van Allen radiation belts, named for physicist James Van Allen, who was in charge of the scientific aspects of the *Explorer* project. While much smaller than the *Sputniks,* weighing only eighteen pounds (8.2 kg), *Explorer* orbited Earth at a higher level, 2,000 miles (3,219 km).

The "space race" was on. Congress, prodded by then Senate Majority Leader Lyndon Johnson, created NASA (National Aeronautics and Space Administration) in July 1958. Over the next few years, the Soviet Union had more successes with their launches than the United States, although some of their missions failed, too. Vanguard rockets proved to be a particular problem, failing again and again. But by the early 1960s, the U.S. effort was moving along well enough so that even the Vanguard failures could be turned into jokes. A famous Mike Nichols and Elaine May sketch centered on a telephone conversation between a mother and her Vanguard engineer son who never seemed to find time to call her. The mother noted in respect to the Vanguards, "you keep losing them," and voiced her concern that their cost would be taken out of her son's paycheck. "A mother worries," she said.

I still remember how awe-inspiring it was to find myself standing in the darkness amid the ruins of the Temple of Apollo at Delphi, Greece, in July 1961, and looking up to see one of the unmanned test vehicles for the Soviet *Soyuz* spacecraft passing across the sky, so high that it reflected sunlight from the other side of the planet. It was my first glimpse of an orbiting satellite, and it seemed doubtful that it could have occurred in a more wondrous or appropriate place.

For advocates of space exploration like Arthur C. Clarke, these were heady days. When the Soviet Union put the first human in space, Yuri Gargarin, who made a single orbit of Earth on April 12, 1961, President John F. Kennedy announced within a month the Apollo program to land men on the moon by the end of the decade. Clarke was already a well-known and popular author, but now he was becoming a prophet in his own time. His 1962 book *Profiles of the Future: An Inquiry into the Limits of the Possible* became a major international best-seller. Books on space exploration would flow from his fertile mind with great regularity over the next dozen years, including another major best-seller in 1968, *The Promise of Space.*

Long before *Apollo 11* actually landed on the moon in July 1969, however, Clarke's 1945 concept of communications satellites in geosynchronous orbits would become a reality. The first step toward such satellites was taken in 1960, when AT&T filed papers with the Federal Communications Commission requesting permission to launch an experimental satellite. The federal government had no policies in place to deal with such a technological development—as would happen again with the Internet two decades later. AT&T was given permission to proceed with its *Telstar*, but the government wanted to make sure that it had its own hand in this field, and NASA awarded a contract to RCA in mid-1961 to build a *Relay* satellite. These were both medium-orbit satellites, at 22,300 miles (36,000 km), below the level of what was already becoming known as the Clarke orbit. NASA also became involved on this front, contracting with the Hughes Aircraft Company to build a satellite of the geosynchronous kind originally suggested by Clarke. This satellite was named *Syncom*. The Defense Department already had a geosynchronous satellite under development, *Advent,* but it was never launched, due to problems with both the Centaur rockets and the satellite itself.

Telstar was the first communications satellite to go up, on July 10, 1962, and provided the first relay of television programs from the United States to Europe. The first geostationary satellite was *Syncom 2,* launched July 26, 1963. As David J. Whalen reports in a NASA history of satellite communications, two Telstars, two *Relays,* and two *Syncoms* had functioned in space by the end of 1964. Then, on April 6, 1965, the Communications Satellite Corporation (COMSAT) launched the first *Early Bird* from Cape Canaveral. That was the start of global communications, although it was not until 1969 that the Intelstat-III series of geosynchronous satellites would provide true global coverage.

In the years since, numerous other kinds of satellites have been launched for special communications needs. Most recently, starting in the late 1990s, Low Earth Orbit (LEO) satellites have been launched by a number of companies for the specific purpose of handling cellular phone traffic, although several of the companies behind LEOs, most famously Irridium, have gone out of business or found themselves in great financial difficulty. The orbit these satellites occupy is only 472 miles (800 km) to 1,491 miles (2,400 km) above Earth. That means that a far greater number of satellites are needed to cover the same territory, or "footprint," as a larger satellite in a Clarke orbit. While LEO satellites are much cheaper to produce, the sheer number needed has involved huge startup costs, and the technology used is more complex. There are also Middle Earth Orbit (MEO) satellites that are placed between the LEOs and those in the classic Clarke orbit.

Communications satellites have changed the world markedly, including the way we think and the manner in which we react to what is happening on the other side of the globe. To give a particularly important example, during World War II the news from the front was carefully monitored by the U.S. government. No picture of a dead American serviceman appeared in a photograph until two years after the war. There were photos of wounded members of the military, but when *Life* printed the first picture of American corpses in 1947, it caused an uproar. During the latter stages of the Vietnam War, however, satellites brought live pictures of battles into our living rooms, a fact that many historians and social commentators believe played a large part in the growing public antipathy to that war. The global village, such as it is, was created by communications satellites.

Arthur C. Clarke has remained an immensely popular writer of both fiction and nonfiction. Later novels such as 1968's *2001: A Space Odyssey* (and its sequels), and 1973's *Rendezvous with Rama* (and its sequels) have been best-sellers. He has won the top science fiction awards, the Hugo and the Nebula, many times. But nothing has brought him more honors than the ideas set down in the short article he wrote in 1945, when he was a twenty-seven-year-old RAF officer with only a high school education. In the years since communications satellites became a reality, he has been awarded a special Emmy Award for engineering for his contributions to satellite broadcasting (1981), a Marconi International Fellowship (1982), the Centennial Medal of the Institute of Electrical and Electronics Engineers (1984), the Charles A. Lindbergh Award (1987), the Lord Perry Award (1992), NASA's Distinguished Public Service Medal (1995), and a host of similar honors. He has also been named to the Society of Satellite Professionals Hall of Fame (1987), and the Aerospace Hall of Fame (1988). And, of course, when he turns on television news, he can be quite sure that some of the pictures he will see will have been transmitted from a satellite traveling in an orbit that bears his name.

As this book was being written, however, Clarke had other things on his mind, particularly the eventual human settlement of Mars. He was certain that would happen, too, on the way to the stars.

To Investigate Further

Clarke, Arthur C. *Greetings, Carbon-Based Bipeds!: Collected Essays 1934–1998.* New York: St. Martin's Press, 1999. This huge collection of the essays Clarke believes worth reprinting is full of classics on innumerable scientific subjects, and is held together by an extremely engaging running commentary. It concludes, as it ought, with an eye-popping set of predictions about the near future, concluding with the words "2100—History begins . . . "

Clarke, Arthur C. *Ascent to Orbit, a Scientific Autobiography: The Technical Writings of Arthur C. Clarke.* New York: Wiley, 1984. Many of these essays also appear in the later book listed above. Quite often, however, the version that appears here (including his 1945 satellite article) contains technical or mathematical material that has been excised from the more popularly oriented 1999 volume.

Clarke, Arthur C. *Astounding Days: A Science Fictional Autobiography.* New York: Bantam, 1989. This book is a must for those interested in Clarke's development as a writer and in science fiction in general. Both this book and *Ascent to Orbit* are most easily obtained from such used book sites as www.alibris.com., but they can also be found in a great many public libraries.

Buderi, Robert. *The Invention That Changed the World.* New York: Simon and Schuster, 1996. Although Clarke himself is a minor player here, this book about the development of radar in Great Britain and the United States just before and during World War II is an enthralling account of a major technological breakthrough, on a par with Richard Rhodes's famous *The Making of the Atomic Bomb.* Rhodes himself says on the book jacket, "The atomic bomb was a sideshow in World War II compared to radar . . . "

CHAPTER 8

Thomas Jefferson

First Modern Archaeologist

The standing of America's Founding Fathers tends to rise and fall with the current political climate in the country. As of this writing, John Adams is very much in vogue, while Thomas Jefferson has come in for a good deal of criticism. But this is a matter of politics, as well as the perpetual jockeying of historians seeking revisionist acclaim. Here, we can largely put politics aside and concentrate on questions of science. In that respect, there is little or no disagreement that Thomas Jefferson and Benjamin Franklin were the primary figures among the men who created our country. Franklin is widely known for his inventions, including the Franklin stove and bifocal eyeglass lenses, and renowned for his experiments with electricity, as we saw in chapter 4. Perhaps because Jefferson was the principal author of the Declaration of Independence, served as governor of Virginia, and was elected both vice president and president of the United States, his scientific interests and work are less well known than they should be. His importance as a politician and statesman is so great that other aspects of his remarkable life tend to be overwhelmed. He is certainly known as a great architect, with his home, Monticello, and the University of Virginia giving ample evidence of those accomplishments. But there was a great deal more to Jefferson than that.

Like Franklin, he was an inventor. He invented the dumb-waiter, designed to bring food from the basement kitchen at Monticello up to the dining room. The dumbwaiter became a fixture in grand houses and hotels around the world, and was eventually electrified in the twentieth century. The first garbage disposal unit was also designed by Jefferson, although it did not take the form we now know. Using a pulley system, as he had for the dumbwaiter, he made it possible to place garbage in a container at the kitchen door that would then travel on ropes to a pit a hundred feet from the house. By pulling on another rope, the container could be tipped over so that it emptied into the pit, after which it made the return trip to the kitchen. The fame of more than a few inventors rests on lesser wonders.

For Jefferson, however, these were just useful gadgets created to make life easier for his slaves—and that was the motivation, more complex racial debates aside. Jefferson's three foremost scientific interests were of a very different kind. He had a greater knowledge of botany than all but a very few people in the United States. His interest in forms of animal life was so great that he set out to disprove a theory of the great French naturalist the Comte de Buffon, amassed a collection of mammoth (or mastodon) bones, and, at the end of his presidency, set up a display of bones brought back from the Lewis and Clark Expedition in the unfinished East Room of the White House. Beyond that, his fascination with Indian burial mounds led him to investigate them in an entirely new way, using a method that has gained him the honor of being called "the father of modern archaeology."

Born on April 13, 1743, into a wealthy Virginia family, Jefferson's education was as comprehensive as that of any man of his time anywhere in the world. He was schooled in Greek, Latin, French, and Spanish, in literature and mathematics, and had knowledge of an astonishing range of scientific subjects. In an article from 1919 in *Natural History,* "Thomas Jefferson's Contributions to Natural History," John S. Patton noted that while it "would not do to proclaim profound scholarship for him in all instances," because of the many other demands on his time, he was an exceptionally eager student of numerous fields, including "mechanics, astronomy, physics, civil engineering (mensuration, strength of materials), surgical anatomy, geology, zoology, botany, economic entomology, aeronautics, and paleontology."

Let's begin with Jefferson's botanical interests. From the age of twenty-three on, he kept his *Garden Book.* Its first entry in 1776, was a simple note: "The Purple hyacinth begins to bloom." Hundreds of pages of entries, some quite lengthy, were added over the years. The final entry in 1824, when he was eighty-one and his health was beginning to fail, was a kitchen garden calendar, including both planting dates and expected harvest times, with a meticulous accounting of the location of all plants. During his trips abroad, particularly when he served as George Washington's secretary of state from 1790 to 1793, he collected specimens of European plants, many of them not indigenous to the United States, and brought them back with him. He also introduced to America plants and shrubs from Africa and China, including the Egyptian Acacias (*Mimosa nilotica*). While his interest in botany was at this stage as much a hobby as anything else, his knowledge was widely recognized. In 1792, Benjamin Smith Barton, a professor of botany and natural history at the University of Pennsylvania in Philadelphia, rose to read a letter to a small gathering of the American Philosophical Society. This letter to a European colleague proposed that an American wildflower, the twinleaf, recently determined to be a distinct genus and therefore in need of a botanical name, be called "Jeffersonia." He had chosen the name not because of Jefferson's political eminence but because of his knowledge of natural history. "In the various departments of this science, but especially in botany and in zoology, the information of this gentleman is equaled by that of few persons in the United States." Barton himself, a native of Lancaster, Pennsylvania, was certainly in a position to judge—in 1803, he would publish *Elements of Botany*, the first textbook on the subject to be written by an American.

As the Thomas Jefferson Center for Historic Plants reports, Jefferson's expertise in matters of natural history was so well known that it had been used against him in the spring of 1792, in an anonymous political pamphlet published in Massachusetts that called for his "speedy retreat" to Monticello, where he could "range the fields and the natural history of his county" instead of inflicting lasting damage to the whole nation as secretary of state. Jefferson replied to this broadside by writing, "However ardently my retirement to my own home and my own affairs may be wished for by others as the author says, there is no one of them who feels the wish once where I do a thousand times."

A small dose of political history is necessary here. Jefferson, who was severely at odds with Secretary of the Treasury Alexander Hamilton, did indeed retreat to Monticello for three years, 1794 to 1796, and busied himself with founding a new party, the Democratic Republicans. Hamilton and John Adams were Federalists, a party that eventually metamorphosed into the Whigs, which in turn fell apart and was replaced in importance by the Republican Party, founded in 1854, while Jefferson's Democratic Republicans became what is now the Democratic Party under Andrew Jackson. In the election of 1796, Jefferson was narrowly defeated by Adams for the presidency, and under the Constitutional rules then in effect, became vice president. Jefferson then defeated Adams in the bitter election of 1800. Aaron Burr was Jefferson's running mate, and nearly became president when the electoral votes for the offices of president and vice president produced a tie, throwing the election into the House of Representatives. The Federalists conspired to make Burr president instead of Jefferson. Adams, furious at his defeat, did nothing to stop them. Alexander Hamilton, however, who was the most influential Federalist leader aside from Adams, trusted Burr even less than he did his old adversary Jefferson, and persuaded enough of his followers to abstain in the House vote so that Jefferson was finally, and rightfully, made president. This mess brought about the Twelfth Amendment in 1804, which made clear that the positions of president and vice president as set forth on the ballot would stand unaffected by the fact that both candidates received the same number of electoral votes. In the midst of the 1801 election crisis, it was noted by a friend, Jefferson seemed much more interested in the news of a new mammoth that had been discovered in upstate New York, which the great portraitist and naturalist Charles Willson Peale (who had painted many of the Founding Fathers) and his son Rembrandt had gone to properly unearth. They kept Jefferson, a close friend, advised of their progress, taking the mind of the might-be president off the political difficulties at hand.

As the second occupant of the White House (Adams moved into the unfinished edifice at the end of his single term), Jefferson doubled the size of the United States with the Louisiana Purchase from Napoleon in 1803. This new territory, ranging from the Gulf of Mexico north almost to what is now Canada, and covering the area between the Mississippi River and the Rocky Mountains, with the exception of Texas, was largely unexplored. Jefferson immediately started to put together the Lewis and Clark Expedition to investigate this new land and to plant the American flag at points along the way. Meriwether Lewis, Jefferson's private secretary and an army captain, was named to head the expedition along with William Clark, whose older brother George had been a notable Revolutionary War general. Jefferson wanted to know everything about the new territories, and along with their primary mapmaking duties, Lewis and Clark were charged with collecting botanical and zoological specimens.

In preparation for the expedition, Lewis was dispatched to Philadelphia to take a cram course in botany from Jefferson's old friend Benjamin Barton. The expedition was supposed to take two years, and when it stretched on into three, it was widely feared that the explorers had perished. When they finally returned to Missouri in 1806, after having traversed the continent all the way to Oregon and back, the country celebrated with an enthusiasm that eclipsed the festivities following the British surrender at Yorktown in 1781. With the help of their remarkable guide Sacajawea, a Shoshone woman who had been captured and sold to a Mandian Indian who then traded her to the expedition's interpreter Toussaint Charbonneau, Lewis and Clark completed one of the greatest exploratory journeys in world history. And to the great pleasure of Thomas Jefferson, they had brought back with them abundant zoological and botanical specimens.

Lewis and Clark transported 226 plant specimens back across the country. Ninety-four proved to be specimens previously unknown to botanists. Their descendants thrive to this day at the Lewis and Clark Herbarium at the Academy of Natural Sciences in Philadelphia, a tangible link to the past that the Herbarium calls the "moon rocks" of 200 years ago. Jefferson, whom Benjamin Barton had honored by naming a wildflower for him, repaid that debt many times over by providing a botanical treasure trove that is still paying scientific dividends. As the Herbarium notes on its web site (www.acnatsci.org/biodiv/botany), in 1997 small pieces of leaf material preserved from the original plants were tested for carbon isotopes to help determine how the atmosphere has changed since preindustrial times.

Botany was a hobby for Jefferson. But in the field of zoology he went beyond that, successfully contesting the view of Georges-Louis Leclerc, Comte de Buffon, that the animals of the New World were degenerate versions of European species. Buffon was enormously influential in his time. His *Histoire Naturelle*, a forty-four-volume work that he began publishing in 1749 (with the final volume appearing in 1804, sixteen years after his death), dealt with all known aspects of natural history. It was a great work and prefigured Darwin by suggesting that some species were extinct, including the American mammoth, bones of which had first been found in 1705. Despite its many virtues, however, it thoroughly annoyed Americans by insisting that animals common to both the Old and New World were smaller in the New, that those indigenous to the New World were smaller in general, and that there were fewer species existing in the New World. There was a hint that Buffon regarded the citizens of the New World as prone to physical degeneracy themselves, which must have seemed a particularly strange notion to Jefferson and Washington (who was also annoyed by Buffon), since they were both over six feet tall and would have towered over Buffon himself.

Jefferson decided to do something about this calumny, and proceeded to write his extraordinary *Notes on the State of Virginia*, published in 1782. The book was far more than a rebuff to Buffon, however, covering not only Virginia's flora and fauna, but also its laws and customs, even containing a meticulous history of the Native American tribes in the region. Jefferson's book was a great success, and was translated into several languages, including French. To back up his assertions about the size of New World animals, Jefferson had friends send him bones of a number of a varieties, including moose, caribou, elk, and deer. On October 1, 1787, he sent the skeleton, dried skin, and antlers of a moose along with the horns of the other animals to Buffon, along with a letter describing them. He concluded by writing, "The Moose is perhaps of a new class. I wish these spoils, Sir, may have the merit of adding anything new to the treasures of nature which have so fortunately come under your observation, & of which she seems to have given you the key: they will in that case be some gratification to you, which it will always be pleasing to me to have procured, having the honor to be with these sentiments of the most perfect esteem & respect, Sir, your most obedient & most humble servant." Such were the niceties of eighteenth-century one-upsmanship.

Buffon, trapped, wrote back to thank Jefferson for enlightening him.

Jefferson already had a wildflower named for him, and would subsequently have an extinct animal named for him. In *The True Thomas Jefferson*, published in 1901, William Elroy Curtis amusingly related part of this story: "In Greenbriar County, Virginia, in 1796, a deposit of bones, supposed to be those of a mammoth, were found and sent to Monticello, where Jefferson set them up and pronounced them to be those of 'a carnivorous clawed animal entirely unknown to science.' A curious sight might have been witnessed by people who lived the route of travel between Monticello and Philadelphia when the vice president of the United States, on his way to take the oath of office and assume the second place in the gift of the nation, carried a wagon-load of bones for his baggage." In Philadelphia, which was still the U.S. capital at that point, Jefferson delivered the bones to the naturalist of the American Philosophical Society, a Dr. Wistar. Jefferson had written a report on the bones, his only true scientific paper, dated March 10, 1797. (Inauguration Day was then in March, because of the difficulty of travel in the winter months—it was changed to January in 1936 for Franklin D. Roosevelt's second inaugural.) The report was titled "A Memoir of the Discovery of Certain Bones of a Unknown Quadruped, of the Clawed Kind, in the Western Part of Virginia." The bones were eventually identified as the first giant sloth to be discovered in North America, and the extinct animal was given the name *Megalonyx Jeffersoni*.

Other presidents have innumerable buildings, streets, and even airports named for them. There are even presidential roses, but those are unofficial, commercial tags. Jefferson alone had the honor of having his name attached to the scientific, Latin description of examples of both the flora and the fauna of the country he had done so much to bring into being. In neither case was this a matter of political tribute, but rather an affirmation of his scientific knowledge.

The botanical and zoological matters we have looked at so far, however, pale beside a true scientific breakthrough made by Jefferson—although his other scientific interests do suggest the remarkable curiosity and the cast of mind that contributed to his greatest success as an amateur scientist. This accomplishment evolved from Jefferson's great interest in Native American tribes. From the beginning, American colonists had been of two minds about the peoples they called the Indians. There is little question that the assistance of Samoset, a secondary chief of the Permaquids, and Squanto, a Pawtuxet who attached himself to the Plymouth Colony, made it possible for the Pilgrims to survive their first two winters, teaching the colonists how to raise and cook indigenous foods. But as more and more colonies sprang up, inevitable conflicts with Native American tribes ensued, with bloody results. Many of the Founding Fathers, even so, were fascinated by Native Americans, and some, like Jefferson and Franklin, learned a great deal about them. In 1754, Franklin proposed that if the colonies were to unite, they might do well to emulate the self-governing tribes of the Iroquois League (the Cayuga, Mohawk, Oneida, Ondonaga, and Seneca tribes), formed in 1570. The Iroquois League was perhaps a trifle advanced for the European settlers, however; everyone had the vote, the women casting not only their own votes but also the proxy votes of their children.

Although there were numerous battles fought with Native American tribes in the seventeenth and eighteenth centuries, there was still a sense in the new nation that many tribes were helpful and trustworthy, and that peace could be worked out with more warlike ones. In 1787, the Continental Congress passed the Northwest Ordinance, which called for Native American rights, the creation of reservations, and the protection of Native American lands. Unfortunately, it also set forth a process for developing what was called the Old Northwest, the area around the Great Lakes. As increasing numbers of settlers moved ever further west, greater conflicts developed, a process that was hastened by Jefferson's own Louisiana Purchase. The tribes of the Great Plains, used to roaming far and wide, were particularly resistant, and decades of fighting would ensue as the nineteenth century wore on. By midcentury, something bordering on genocide was taking place. In the years 1853–56, the government of the United States signed fifty-two treaties with Indian tribes. In the end, the government broke every single one of them.

But that dark period was not foreseen in Jefferson's time. In his *Notes on the State of Virginia*, Jefferson listed the names and original locations of dozens of tribes in the colonies and known territories, using elaborate tables that included population figures where he could obtain them. This "Catalogue," as he called it, was drawn from four different lists put together by traders and Indian agents. It is a remarkable document of great historical importance. That catalogue was followed by an account of his investigation of an Indian barrow, or mound, situated "on the low grounds of the Rivianna, about two miles above its principal fork, and opposite to some hills, on which had been an Indian town."

As an introduction to his investigation of the mound, Jefferson noted that there were four theories about their purpose and construction.

That they were repositories of the dead, has been obvious to all: but on what particular occasion constructed, was a matter of doubt. Some have thought they covered the bones of those who have fallen in battles fought on the spot of internment. Some ascribe them to the custom, said to prevail among the Indians, of collecting, at certain periods, the bones of all their dead, wheresoever deposited at the time of death. Others again supposed them the general sepulchres for towns, conjectured to have been on or near these grounds; and this opinion was supported by the quality of the lands in which they are found, (those constructed of earth being generally in the softest and most fertile meadow-grounds on river sides) and by a tradition, said to be handed down from the Aboriginal Indians, that, when they settled in a town, the first person who died was placed erect, and earth put about him, so as to cover and support him; that when another died, a narrow passage was dug to the first, the second reclined against him, and the cover of earth replaced, and so on. There being one of these in my neighborhood, I wished to satisfy myself whether any, and which of these opinions were just.

This splendid elucidation of purpose is typical of Enlightenment thought, remarkably similar in tone and structure to the works of Joseph Priestley discussed in chapter 4—it is hardly surprising that Priestley and Jefferson should have become friends when the former emigrated to America. The phrase "I wished to satisfy myself . . . " could indeed be the credo of the amateur scientist down through the ages. Conflicting opinions existed. Which of them was "just"? The author of the Declaration of Independence would of course use that word. But if Jefferson's way of expressing why he undertook this project is typical of the Enlightenment, the way he would go about satisfying himself was revolutionary. No one had ever done it before.

Up to this point in history, archaeology was far from being the science it would become. Much of the "investigation" of earlier civilizations had not risen much above looting, and some of it was exactly that. There are beautiful objects from antiquity in museums all over the world that were torn from the ground, or from tombs, in a "get-in-and-get-out" haste that destroyed most of the evidence modern archaeologists use for dating and providing a larger historical context. Even those with a more scholarly attitude, who tried to work slowly enough to avoid undue damage to buried treasure, used haphazard methods that often obliterated vital clues to the real meaning of many precious objects. Bones, human bones in particular, were likely to be treated with little respect. By the eighteenth century, animal bones, if they were large enough—like those mammoths—were treated with greater care. It was beginning to be recognized that the ground was full of bones—like Jefferson's giant sloth—that were extremely mysterious, and it became obvious that the more pieces of a skeleton could be recovered, the greater the possibility of solving the mystery.

Newton had begun to establish modern scientific method for physics with his experiments on light in 1665, but as we have seen in this book, it was not until the mid-eighteenth century that experimenters like Priestley began to use modern methods in respect to chemistry, often inventing the techniques as they went along. One reason that archaeology lagged behind was the prevailing belief that the Earth was only about 6,000 years old, created by the biblical God in His image. Buffon, despite his disparagement of New World animals, was way ahead of his time in suggesting that such creatures as the mammoth might be extinct. Jefferson himself held out hope that Lewis and Clark might find living mammoths on their trek westward, although this was not necessarily a result of religious piety. The religious views of the Founding Fathers tended to be in flux to some degree, which is one reason why commentators in our own time can find passages to prove whatever beliefs suit their own dogma. Jefferson, for example, suggested in one of his letters to Joseph Priestley that the day might come when all young men might be Unitarians—a heretical statement in the context of the times. Jefferson was not quite willing to embrace Unitarianism for himself, but as an open-minded son of the Enlightenment, he could see that such a possibility existed. Most others had difficulty in going even that far, however. Thus the scientific aspects of archaeology that are now taken for granted could be seen as unnecessary to most investigators. If Earth was only 6,000 years old, "ancient" still meant something quite recent, an attitude that slowed the development of methodologies for determining such dates.

Jefferson, however, in line with the meticulous notes in his *Garden Book,* approached his investigation of the mound with intellectual rigor. The mound was not undisturbed even as it stood. Some forty feet in circumference, it was about seven and a half feet high. It had previously been about twelve feet high, he knew, and covered with substantial trees. The spheroid had been leveled off and the trees cut down a dozen years earlier, and the top of the mound was under cultivation. There was an ancient trench some five feet wide and deep around the entire perimeter, from which earth had clearly been removed to build the mound in the first place.

Jefferson first dug superficially at various points on the mound, as anyone else at the time would have done. He came upon

collections of human bones, at different depths, from six inches to three feet below the surface. These were lying in the utmost confusion, some vertical, some oblique, some horizontal, and directed to almost every point of the compass, entangled, and held together in clusters by the earth. Bones of the most distant parts were found together, as, for instance, the small bones of the foot in the hollow of a scull [*sic*], many sculls would sometimes be in contact, lying on the face, on the side, on the back, top or bottom, so as, on the whole, to give the idea of bones emptied promiscuously from a bag or basket, and covered over with earth, without any attention to their order.

Helter-skelter though the bones were, Jefferson examined them with care.

The sculls [*sic*] were so tender, that they generally fell to pieces on being touched. The other bones were stronger. There were some teeth which were judged to be smaller than those of an adult; a scull, which, on a slight view, appeared to be that of an infant, but it fell to pieces on being taken out, so as to prevent satisfactory examination; a rib, and a fragment of the under-jaw of a person about half grown; another rib of an infant; and part of the jaw of a child, which had not yet cut its teeth. This last furnishing the most decisive proof of the burial of children here, I was particular in my attention to it. It was part of the right-half of the under-jaw. The processes by which it was articulated to the temporal bones, were entire; and the bone itself firm to where it had been broken off, which, as nearly I could judge, was about the place of the eye-tooth. Its upper edge, wherein would have been placed the sockets of the teeth, was perfectly smooth. Measuring it with that of an adult, by placing the hinder processes together, its broken end extended to the penultimate grinder of the adult. The bone was white, all the others of a sand color. The bones of infants being soft, they probably decay sooner, which might be the cause so few were found here.

The forensic detail of this passage is in itself highly unusual for the time. In the Old World, the ruins of Herculaneum, below the volcano Mt. Vesuvius, had been discovered in 1709. The Prince of Elbouef uncovered the first complete Roman theater, but was far more interested in unearthing works of art for his collection, and did not bother to record where any of them were found. Pompeii was discovered in 1748, with much the same results, as the King and Queen of Naples added to their collection. Indeed, it wouldn't be until 1860, as reported by Colin Renfrew and Paul Bahn in their highly regarded book *Archeology*, that well-recorded excavations began to be carried out by Giuseppe Fiorelli. The kind of detail provided by Jefferson was far ahead of its time.

Jefferson's next step was quite simply revolutionary. He made a perpendicular cut to within about three feet of the center of the mound. It was wide enough for Jefferson to walk into in order to examine the sides of the cut. That revealed the various strata of the mound from bottom to top.

At the bottom, that is, on the level of the circumadjacent plain, I found bones; above these a few stones, brought from a cliff a quarter of a mile off, and from the river one-eighth of a mile off; then a large interval of earth, then a stratum of bones, and so on. At one end of the section were four strata of bones clearly distinguishable; at the other, three; the strata in one part not ranging with those in another. The bones nearest the surface were the least decayed. No holes were discovered in any of them, as if made with bullets, arrows or other weapons. I conjectured that in this barrow might have been a thousand skeletons.

The concept of examining the strata of a mound, or of anything else, was entirely new. It would not be until the following year that the Scottish geologist James Hutton would publish his *Theory of the Earth.* As Renfrew and Bahn succinctly put it, Hutton "had studied the stratification of rocks (their arrangement in superimposed layers or strata), establishing principles which were to be the basis of archeological excavation, as foreshadowed by Jefferson."

Jefferson was able to draw some very specific conclusions from his new approach to archaeological excavation:

Every one will readily seize the circumstances above related, which militate against the opinion, that it covered the bones only of persons fallen in battle; and against the tradition also, which would make it the common sephulchre of a town, in which bodies were placed upright, and touching each other. Appearances certainly indicate that it has derived both origin and growth from the accustomary collection of bones, and deposition of them together; that the first collection had been deposited on the common surface of the earth and a few stones put over it; and then a covering of earth, that the second had been laid on this, had covered more or less of it in proportion to the number of bones, and was then also covered with earth; and so on. The following are the particular circumstances which give it this aspect. 1. The number of bones. 2. Their confused position. 3. Their being in different strata. 4. The strata in one part having no correspondence with those in another. 5. The different states of decay in these strata, which seem to indicate a difference in the time of inhumation. 6. The existence of infant bones among them.

In 1784, when Jefferson made this archaeological survey, the first of its kind anywhere in the world, mounds like the one he excavated were a great mystery. Contemporary Native American tribes were building nothing like them, and could not explain them, either. Some scholars suggested that they belonged to a far earlier civilization. It would not be until the 1960s that work by archaeologists, anthropologists, and historians managed to arrive at a consensus about their origins. Mounds were to be found in many parts of the country, from the Southeast through the Mississippi Valley into the Midwest. This suggested that there could have been more than one group of mound builders, and in the end it was concluded that there were three different cultures that had created them. The oldest of these cultures, called the Adena after an Ohio farm that contained a particularly large mound, flourished from 1000 B.C.E. to about A.D. 200 in the Ohio Valley. A second culture overlapped somewhat with the first, existing from 300 B.C.E. to A.D. 700. This Hopewell culture spread eastward out of the Ohio Valley, and it was a Hopewell mound that Jefferson investigated. A third culture, the Temple Mound Building (or Mississippian), began around 700 A.D., and its final vestiges still existed in the seventeenth century, although its people had stopped constructing mounds.

Jefferson himself speculated in *Notes on the State of Virginia* that the original peoples of the Americas might have come from Asia, on the basis of the fact that Captain Cook had proved that the Asian and American continents were either connected, or separated by a narrow strait, in the far north. We now know that the Bering Strait was indeed a land bridge between the two continents until about 30,000 years ago, and it has been well established that the Native American population did indeed originate in Asia. Even in such speculative matters, the depth of Jefferson's knowledge, combined with his extraordinary intelligence, allowed him to make some good guesses.

Although Jefferson's account of his mound excavation is now judged one of the most extraordinary passages in *Notes on the State of Virginia*, the new archaeological method he used did not quickly become a standard one. He was decades ahead of his time in this regard. It would be nearly a hundred years before stratigraphic excavations began to become the norm. During the nineteenth century, the discovery of dinosaur bones, evidence that there had been ice ages in the past, and Darwin's theory of evolution gradually changed the way human beings thought of Earth and its history. Only when it began to be understood that Earth had existed for millions instead of mere thousands of years did the need for extremely careful archaeological excavation become fully apparent. Once that approach became standard, it was soon recognized that Thomas Jefferson had shown how an excavation ought to be carried out way back in 1784.

Some men and women of genius excel in fairly narrow fields of endeavor. That does not lessen their accomplishments or the eventual impact of their ideas—Henrietta Swan Leavitt's recognition of the importance of Cepheid stars, as we saw in chapter 3, led to a new understanding of the entire universe. Thomas Jefferson's political genius helped define a democratic ideal that has changed the world. His scientific explorations in several fields did not carry that kind of weight. But they do help explain why, like Leonardo da Vinci or Sir Walter Raleigh, he is regarded as being a Renaissance man, an individual with a breadth of interests and talents that even today can create a sense of awe. At a White House dinner that included many of the finest minds in American, President John F. Kennedy rose to make a toast and, looking around at the assembled notables, said that they probably constituted the greatest concentration of intellect in the White House "since Thomas Jefferson dined alone."

To Investigate Further

Jefferson, Thomas. *Writings*. Merrill D. Peterson, ed. New York: The Library of America, distributed by Viking Press, 1984. The publications of The Library of America are invaluable collections of major texts by a wide variety of American writers. This volume includes not only the full text of Jefferson's *Notes on the State of Virginia*, but numerous public papers, his inaugural addresses, and several hundred pages of his endlessly fascinating letters to many of the greatest figures of his time, both American and European. A paperback edition containing only the *Notes*, with an introduction and annotation by Frank Shuffleton, is available from Penguin.

Renfrew, Colin, and Paul Bahn. *Archeology: Theories, Methods and Practice*. New York, Thames and Hudson, 2000. This is the third updated edition of a widely praised book originally published in 1991. Although it is technically a textbook, it is extremely readable and contains more than 600 illustrations.

Ambrose, Stephen E. *Undaunted Courage: Meriwether Lewis, Thomas Jefferson, and the Opening of the American West*. New York: Simon & Schuster, 1996. A bestseller, this account of the Lewis and Clark expedition is written with Ambrose's usual combination of detail and page-turning sweep.

Cohen, I. Bernard. *Science and the Founding Fathers: Science in the Political Thought of Jefferson, Franklin, Adams, and Madison*. New York: Norton, 1995. A fascinating examination of the connection between the burgeoning of science and democratic thought.

Jefferson, Thomas. *The Garden and Farm Books of Thomas Jefferson*. Robert C. Baron, and Henry S. Commager, eds. New York: Fulcrum, 1998. Anyone who loves gardening will find endless delight in these diaries.

Note: In writing this chapter I drew on two sources to which I have a personal connection. The first is the Pulitzer Prize-winning six-volume biography *Jefferson and His Time*, written by my late uncle, Dumas Malone, and published in both hardcover and paperback editions by Little, Brown. I also referred to my own book, *The Native American History Quiz Book*, published in 1994 by Quill/Morrow.

CHAPTER 9

Susan Hendrickson

Dinosaur Hunter

It was August 12, 1990. Sue Hendrickson had hiked three miles across the rocky terrain of north central South Dakota to get a good look at a barren sandstone cliff she had noticed from a distance two weeks earlier. That morning, the small crew from the Black Hills Institute, a private fossil-collecting enterprise, had started to excavate a partial skull of a *triceratops* (one of the best-known dinosaurs, with a three-horned face). Then it was discovered that the battered collecting truck had a flat tire, and the spare was no good, either. Peter Larson, the head of the institute, had decided to make the forty-five-minute drive to the town of Faith in another vehicle to get the tires repaired. He had asked Hendrickson to come along. They had been romantically involved, but had mutually agreed to split up. That might have been a reason not to go with Larson, but the main one was that she saw an opportunity to investigate that cliff in the distance. She said later that it had been "beckoning" to her. "I felt drawn to that formation for two weeks. I can't explain it." It wasn't that Sue had a sixth sense, exactly, but she knew from long experience that it paid to follow her hunches. So she set off with her Labrador retriever, Gypsy, leaving the rest of the group to work on the *triceratops*.

181

It had been foggy that morning, which almost never happened during the summer in that part of the country, and Sue had found herself going in a circle. Once the fog lifted, she set out again. She finally reached the sixty-foot-high wind-carved cliff two hours later. When dealing with a cliff face, the first step that fossil hunters take is to walk along the base of it, their eyes on the ground, looking for "dribbled-down" bones, as Hendrickson puts it. This can be a clue that something is buried in the cliff face that could be worth investigating further. While many people might think it would make more sense to scan the cliff face, the bones of a dinosaur can be hard to spot, especially in sandstone, which weathers into peculiar shapes all on its own. As Sue walked along, her keen eye caught some small pieces of bone lying on the ground. She picked them up, and cast her eye up the cliff face to see where they might have come from. She saw it immediately, about eight feet off the ground. Weathering out of the cliff face were three large vertebrae and a femur. Their size and shape immediately excited her. As she explained to Steve Fiffer, author of *Tyrannosaurus Sue*, "The carnivores like *T. rex* had concave vertebrae from the disk, it dips in. The herbivores—the *triceratops* or duck-bills is what you almost always find—have very straight vertebrae. So I knew it was a carnivore. I knew it was really big. And therefore I felt it must be a *T. rex*, but it can't be a *T. rex* because you don't find *T. rex*."

Only eleven previous *T. rex* skeletons, none complete, had ever been found. The first was dug up in 1902, a second in 1905, and a third in 1907, all discovered by perhaps the most famous of all dinosaur fossil hunters, Barnum Brown. No more were found until 1966. The first was discovered in the Hell's Creek Formation in the Badlands of eastern Montana. Brown, who was working on a commission from the American Museum of Natural History in New York, wrote to Henry Fairfield Osborn, the curator of the paleontology department (and eventual president of the museum): "Quarry #1 contains the femur, pubes, part of the humerus, three vertebrate, and two indeterminate bones of a large carnivorous dinosaur, not described by Marsh. I have never seen anything like it from the Cretaceous." The letter was dated August 12, 1902—eighty-eight years before Hendrickson found her *T. rex*, to the very day.

Public fascination with dinosaurs had its beginnings in the 1840s. Dinosaur fossils had been found earlier, but they were not identified as such. The Lewis and Clark expedition (see chapter 8) came across a gigantic bone near what is now Billings, Montana, in 1804. It was too large, and too difficult to extricate, for them to bring back with them, but their descriptions of it convinced later experts that it must have belonged to a dinosaur. In the early 1820s, Gideon and Mary Anne Mantell found fragments of what would come to be called an *iguanadon* in England, where William Buckland also uncovered the bones of a *megalosaurus* in 1824. In America, the Rev. Edward Hitchcock, the president of Amherst College, discovered fossilized tracks of what he at first believed to be giant birds on the banks of the Connecticut River near Mount Holyoke, Massachusetts. The Footprint Preserve established there can still be visited.

It was not until 1841, however, that the word *dinosaur*, meaning "terrible lizard," was coined by the British anatomist Richard Owen in a two-and-a-half-hour lecture called "Report on British Fossil Reptiles." As increasing numbers of dinosaur fossils were found, new names for and descriptions of these extinct creatures were quickly developed. But because scientists were dealing with fragments in almost all cases, the results were often far off the mark. In 1851, the first World's Fair was held at England's Crystal Palace, erected in Hyde Park in the center of London. Three years later the Crystal Palace was dismantled and reassembled in the suburb of Sydenham. Prince Albert suggested creating replicas of British animals around the site, including examples of dinosaurs. Richard Owen was put in charge of the dinosaur project and hired a sculptor named Benjamin Waterhouse Hawkins to create examples of the *megalosaurus,* the *hylaeosaurus,* and the *iguanadon.* They were all wrong, scientifically (the *iguanadon* looked like a modern iguana enlarged to the size of a rhinoceros), but they were very large and very exotic, and they created a sensation. The first dinosaur rage was born.

The action quickly moved to America, where the variety of dinosaur fossils was far greater than in Europe. The first *hadrosaurus* was found in a marl pit in Haddonfield, New Jersey, by William Parker Foulke, and soon named by Joseph Leidy of the Academy of Natural Sciences in Philadelphia. All of the Hawkins models at the Crystal Palace were depicted as four-footed, which was correct only for the *hylaeosaurus.* The *hadrosaurus* had such short front legs that it was obvious it must have walked upright, and it was soon realized this was true for a large number of dinosaur species. Representations of dinosaurs began to change. Hawkins's later sculptures and drawings, for example, are far more accurate than his Crystal Palace examples.

During the 1860s, two American experts on dinosaurs emerged as preeminent, Edward Drinker Cope and Othniel Charles Marsh. Both came from wealthy backgrounds, and initially they were friends, but by 1877 they had become bitter rivals, sniping at each other's work and competing to amass the greater number of finds. Their exploits as well as the charges of incompetence they flung at each other made headlines. Marsh was head of the Peabody Museum at Yale, from which he dispatched fossil-hunting expeditions and then took credit for their discoveries, while Cope spent much more time doing actual field work. Both men tended to go overboard in naming each new find a different species—a problem that would take decades to untangle—but their rivalry also stimulated the search for American dinosaur fossils as well as informed the debate over evolutionary theory. The public came to expect natural history museums to have major dinosaur displays. Because New York's American Museum of Natural History lagged in this regard, it hired Marsh's protégé, Henry Osborn. Osborn shared his mentor's weakness for seeing a new species at every turn, but he turned the museum into one of the greatest repositories of dinosaur examples, in large part due to the field work of Barnum Brown.

Brown, who had been named for the epic showman P. T. Barnum, dropped his pursuit of a Ph.D. at Columbia University in order to do the field work he loved, working for Osborn beginning in 1897. Like Susan Hendrickson a century later, he seemed to have a sixth sense about where dinosaur fossils might lurk—some went so far as to suggest that he could smell dinosaur remains. He was famously well dressed, even when working in the field, and became one of the most sought-after dinner guests of the era, a man with impeccable manners who was also a dinosaur hunter in the wilds of Montana and Wyoming—what more could a hostess ask? He was famous before he found the first three *T. rex* skeletons, each more complete than the last, but those discoveries made him something of a legend. He would die in 1963 at the age of 100, three years before a fourth *T. rex* was finally found.

Against this background, Susan Hendrickson's excitement on the morning of August 12, 1990, is entirely understandable. To discover a *T. rex* was the dream of every dinosaur hunter, the equivalent of a geologist stumbling onto a diamond mine. The *tyrannosaurus* family, of which the *rex* was the largest and most ferocious in appearance, existed for only 20 million years, at the end of the Cretaceous period, from 85 to 65 million years ago, when the mass extinction occurred that wiped out all the dinosaurs (except the avian dinosaurs, from which birds evolved). The *T. rex*, king of this family, probably did not appear until about 67 million years ago, making it one of the final dinosaurs on Earth. *Triceratops,* the herbivores from the Cretaceous with seven-foot jaws that served as huge threshing machines and triple facial horns that could protect them from predators like the *T. rex,* were not only around for longer but were clearly a much more common animal. The rarity of *T. rex* specimens reflects their place in the ecological spectrum. This can be understood in terms of the numbers of animals in our own time. It takes vast herds of grass-eating antelope to support a single pride of lions, for example.

In addition, the largest animals are least likely to become fossilized in anything close to their entirety. As Michael Novacek puts it in *Dinosaurs of the Flaming Cliffs*, his account of the 1993 expedition from the American Museum of Natural History to the Gobi Desert, which he led, "There is no reason to expect that a skeleton will endure the ravages of any step in its fossilization. It could be shattered and crushed to unrecognizable powder by a suite of scavengers, like those bone-crushing, marrow-sucking hyenas on the Serengeti plain. The moving streams might feed rivers of such power that the bones are broken, beveled, or rounded into indistinguishable pebbles and carried to the sea. The forces of sedimentary compression might squeeze the bones like toothpaste into ghostly smears of their former sharp-edged solidarity." Even those fossils that do survive may be buried deep in the earth, and the shifting forces that could lift them up again to the surface may just as easily destroy them. Once they reach the surface, they may be eroded to dust long before paleontologists find them.

Sue Hendrickson was very lucky. The *T. rex* she found in 1990 had survived largely intact the punishing processes of 65 million years and had reappeared in a cliff on the surface, and the cliff had eroded just enough to reveal a few telltale bones of the great creature, without exposing so much of the skeleton that it would be obliterated. Hendrickson immediately recognized that because the vertebrae were going into the hill, the possibility existed that much more of the skeleton was embedded in the cliff. She took some of the small pieces of bone that were lying at the base of the cliff, none more than two inches across, back to camp to show them to Peter Larson. As she had already noted, they were hollow, like the bones of birds. Most dinosaur bones are solid, and the hollowness was another sign that she had found a *T. rex*. Larson immediately agreed with her. They raced back to the cliff, where Hendrickson told him it was her "going-away" gift to him. As he examined the cliff face, he was quickly convinced that not only had Hendrickson found a *T. rex* but that it must be a nearly complete specimen. It was the find of a lifetime.

———

Sue Hendrickson has often been compared to Indiana Jones, the fictional movie hero portrayed by Harrison Ford in three Steven Spielberg movies. Not only has she had adventures all over the world, but she was born in Indiana in 1949, growing up in the small town of Munster. Her mother had been a schoolteacher, and Hendrickson devoured books. She was the kind of kid who was reading Dostoyevsky at eleven. Like many such kids, she was bored by school and was a rebellious teenager. As Fiffer reports in *Tyrannosaurus Sue*, the differences with her parents became so acute that they decided she might do better to finish high school in Fort Lauderdale, Florida, where an aunt and uncle lived. But that didn't work well, either, and she ran off with her boyfriend. They moved all over the country, taking odd jobs, and finally settled down in San Rafael, California, painting boats for a living. Hendrickson kept in touch with her parents, and even visited them—there was no real break. After splitting up with her boyfriend, she went to Florida and worked in Key West, diving to catch aquarium specimens. As was her usual habit, she learned a great deal about this new endeavor. She knew what she was looking for in the ocean waters, what was wanted. But sometimes she would come to the surface with a specimen that didn't seem to be in any books, or known to anyone else she was working with. When that happened she'd drive the ninety miles to an oceanographic laboratory in Miami.

Sometimes the specimen she'd brought with her was just rare. But on several occasions she brought in a previously unidentified species. With her long blonde hair, Hendrickson looked like a mermaid, and she seemed to have a mysterious ability to spot any unusual creature swimming in the sea. The oceans of the world, like the Amazon basin, still harbor untold numbers of unidentified species. The professionals were impressed. Sue Hendrickson seemed to find more than her share of marvels.

Thinking she might be ready to settle down and go to college, she joined her parents in Seattle, where they had moved, and passed her high school equivalency tests. But the prospect of getting a Ph.D. just to dissect fish changed her mind, and a year later she returned to Florida, where she ended up working as a salvage diver. She also dove for lobsters to sell to restaurants, once hauling in nearly 500 in a single day. Her proficiency as a diver—and her knack for finding things underwater—led to a stint with marine archaeologists in the Dominican Republic. The Dominican Republic has a great many mineral deposits, and Hendrickson realized that the miners often found prehistoric insects preserved in amber in the mountain caves. Amber is a translucent yellow to brown fossil resin originally secreted by an ancestor of the pine tree millions of years ago. Sometimes insects or plant materials were trapped in the resin when it was soft, and are as clear and complete as items we now encase in man-made lucite. Sue Hendrickson did some prospecting for amber on her own, but found it more sensible to buy insects in amber from miners, reselling them to collectors and museums. Once again, she learned a new discipline. While it was easy enough to judge which specimens would attract general collectors because of their beauty, it was important to know that some ugly insect could be of evolutionary significance and thus a museum piece. She made a profit on the pieces she sold to private collectors, but if she procured a piece that was of museum quality, she would sell it at cost. Among her greatest treasures were three complete butterflies—only six examples exist in the world.

Many important museum curators have attested to Hendrickson's generosity, her lack of interest in money or fame, her genuine fascination with nature's bounty, and her desire to share it with the world. Ever curious and anxious to learn about new fields, she found herself digging for the fossils of prehistoric whales in Peru with the Swiss paleontologist Kirby Siber. It was there that she met Peter Larson in 1985. What Hendrickson refers to as her "itchy feet" came to rest for a time. Larson, born in 1952, was a dinosaur hunter whose Black Hills Institute had become a major supplier of specimens to museums around the world, including nine duck-billed dinosaurs during the 1980s, the last three of which would bring more than $300,000 each. Larson was hardly getting rich, however; the excavation of dinosaur bones, and their subsequent preparation for mounting, is a lengthy and expensive process. Hendrickson became romantically involved with Larson, but also learned an enormous amount from him. When it came to new information about any aspect of the natural world, she was a sponge, absorbing as much as anyone could tell her. As always, she proved adept at her new vocation, but it wasn't until she followed her instincts and set off toward the sandstone cliff that she fully demonstrated what an extraordinary "nose" she had. Larson named the creature "Tyrannosaurus Sue" as soon as he saw what they had. Was "Sue" actually a female? There is still debate about that. The size of the skeleton, thirteen feet high at the hip and forty-one feet long, was even bigger than Barnum Brown's best example, at the American Museum of Natural History, by a foot or two in either direction, but that doesn't necessarily indicate a male.

Partially because there are so few *T. rex* fossils, there is a lack of certainty about the habits of these dinosaurs, and every new find raises hopes that some of these problems can be settled. One main area of controversy concerning the *T. rex* has focused on the question of whether they were true predators or more usually scavengers. The idea of a creature with a four-foot-long skull implanted with six-inch serrated teeth being a mere scavenger seems absurd to many scientists—why would they evolve such enormous incisors unless they were predators? But other experts point to the puny forelimbs of the tyrannosaurids in general, which are so short that the animals couldn't even have reached up to their own mouths. The claws on these forelimbs have only two fingers, which means they could not have been used for grasping an enemy. How could they have succeeded in combat with an armored three-horned *triceratops,* these experts ask, even though the latter creature would only have come up to the knee of a *T. rex?* Yet there is fossil evidence, including that from "Sue," that they did engage in such combat. They have the wounds to show it. Michael Novacek writes, "Certainly snakes, crocodiles and other infamous killers have attained this status without the benefits of grasping forelimbs. Indeed, the scimitar-toothed jaws of tyrannosaurids were sufficient to snatch, crush and carry away a victim, or at least carry away a good chunk of meat from the prey." Pronouncing the scavenger theory "a bit overwrought," Novacek suggests that they may have been both predators and scavengers, but that there is no evidence to prove that they weren't "terrific killers." Certainly, "Sue" gave every appearance of having been a terrifying creature, as Hendrickson, Larson, and his crew began to excavate the skeleton. Larson, for one, was utterly convinced, following his first look at "Sue's" jaw, that she was a predator: "Her mouth was her arsenal," he said.

Removing fossils, especially large ones, is never easy, but the degree of difficulty depends on the nature of the site where they are found. In their book *Discovering Dinosaurs*, Mark A. Norell, Eugene S. Gaffney, and Lowell Dingus of the American Museum of Natural History explain the process: "Before the digging starts, a photograph is taken to record the discovery and its location. Then a coating of hardener is applied to all of the exposed bone surfaces to stabilize the specimen. The excavation process depends on the type of matrix in which the specimen is entombed. If the matrix is soft (poorly consolidated sand, or silt), excavation commences with brushes and awls. If the matrix is slightly harder, geologists' picks and shovels are used. If the matrix is very hard (such as cemented sandstone), heavier tools such as jackhammers, rock saws, or even dynamite, may be necessary."

The excavation of "Sue" took the middle approach. The team made videotapes as well as took photographs. All fossil scraps from the ground were collected, bagged, and labeled individually. Bags of dirt were also collected, as is always done at such a site. Future screening of the dirt, including microscopic examination, could reveal fossil remains of other creatures that would add to our knowledge about the end of the Cretaceous period. It is known, for example, that the first mammals had appeared during this time, but they were mostly tiny creatures. Even a fossil the size of a grain of rice can add crucial information about the emergence of mammalian species.

Larson's brother Neal, who had left the camp a week earlier, returned. His fifteen-year-old son Mathew and Peter Larson's ten-year-old son Jason had been with the team all summer. Terry Wentz, who was the head fossil preparer for the Black Hills Institute, would now play a newly important role. The bones that were sticking out of the cliffside were coated with hardener and covered with burlap and plaster of paris. Barnum Brown, according to *Discovering Dinosaurs*, was the first to use this technique on dinosaur fossils in the 1890s. "For five days," Fiffer writes in *Tyrannosaurus Sue,* "the Larson brothers, Hendrickson, and Wentz worked with pick and shovel to clear the sandstone and hard soil above the skeleton. 'These were the hottest days of summer,' says Hendrickson. 'The temperature was 115 plus. You're trying to find shade but there is none. And we don't stop at noon.'"

Hendrickson wasn't really complaining. She would later tell an interviewer, "It was like I was a sculptor—the feeling that you are creating her from the rock, almost bringing her to life." She had been right in her assumption that the *T. rex* extended back into the cliff, and would be nearly complete. In the end, they recovered 90 percent of the skeleton, making "Sue" both the largest and the most complete example ever found. The job actually turned out to be less difficult than many such excavations, in part because the skeleton was so complete and not scattered over a larger area. The greatest difficulty was encountered when it was discovered that "Sue's" skull was under her hips, which weighed 1,500 pounds. It proved necessary to take this area out as a single block in order to avoid crushing the skull in the process.

Eighty-eight years earlier, when Barnum Brown discovered the first *T. rex* in Montana, he had faced far more formidable problems. There were more pieces to excavate, and they had to be hauled to the nearest railway, which was 125 miles away. The process took months. "Sue," on the other hand, had been freed from the rock that had imprisoned her for millions of years in only seventeen days. Mission accomplished, Hendrickson headed off for new adventures. She was thrilled to have found a *T. rex*, but she had itchy feet again. Nor was she all that happy about having the dinosaur named for her. She wasn't really interested in that kind of publicity, for one thing. For another, she had never liked her name. She regarded "Sue" as a little better than "Susan," but not much. Somehow, it didn't sound like the name of an explorer.

When she left South Dakota, Hendrickson had no idea how famous her find would make her—or how peculiar the circumstances under which she'd eventually return to South Dakota would be.

———

Peter and Neal Larson did not have advanced degrees in paleontology, but they had been professional fossil hunters for a long time, Peter since 1974, Neal since 1977. They knew the ropes, not only in the field but in terms of the museum world. They certainly knew that you couldn't just go dig anywhere you wanted. You had to get permission from the owners of the land, whether that owner was a private individual, the state, or the federal government. That summer, they had permission to hunt in an area known as Mason's Quarry, a famous site for dinosaur finds. The land they found Tyrannosaurus Sue on, however, belonged to a rancher named Maurice Williams, who was one-quarter Native American. His ranch, a large one, lay on the Cheyenne River Sioux Reservation. Earlier in the summer the Larsons had found a dead horse that turned out to belong to Williams. He was very interested in what they were doing and invited them to explore his own land. Peter Larson told him that if they found anything, they'd pay Williams, although he cautioned that they didn't pay a lot. That was fine by Williams, and by his brother Sharkey, who invited the team to hunt for fossils on his land, too.

As Fiffer reports, they didn't find anything on Maurice's land, but Sharkey's did produce a few partial *triceratops* skulls, one of which they were excavating the morning Susan Hendrickson hiked off to the beckoning cliff in the distance. That cliff was on Maurice's land, and once they'd finished digging the huge fossil out, Larson signed a contract with him and gave him a check for $5,000. That, of course, is a very small sum for such an important find, which might bring a million dollars from a museum. Larson wasn't planning to sell it, however. He saw it as the cornerstone of the museum he had always wanted to build in connection with the Black Hills Institute in Hill City. Indeed, that is exactly what he announced at the annual meeting of the Society of Vertebrate Paleontologists, held in San Diego in October 1991. He even invited the assembled paleontologists to come to the Black Hills Institute to conduct their own investigations of Sue Hendrickson's highly important find.

The remains of three smaller *T. Rex* had been found with "Sue." They were clearly juveniles, and to Larson that suggested he had been right in his hunch that the dinosaur was indeed a female. Somehow, she had been killed with her family. He embarked on a project to see what the differences between the sexes were in other dinosaur species of which far more examples had been found, and whose sex has been determined. In many cases the females were the larger, so the very size of "Sue" could in fact indicate that she was a female. Among humans, the male is generally larger, but there are many exceptions to that rule among other species.

In addition, Larson hoped that "Sue" would shed some light on the long, bitter debate about whether dinosaurs were warm-blooded or cold-blooded animals. There were also questions about brain size to be dealt with. NASA was persuaded to carry out a CAT scan of the skull, and preparations were made to ship it to Huntsville, Alabama, on May 17, 1992. But on April 29, the *Rapid City Journal* carried a headline revealing that the Sioux tribe Maurice Williams was associated with had filed suit to regain "Sue." On May 14, the FBI, backed up by the National Guard, raided the Black Hills Institute and prepared to haul "Sue" away.

The resulting mess is too complex to report on in any detail here. Steve Fiffer spends more than half his book, *Tyrannosaurus Sue*, dealing with the ins and outs of this much-publicized case, which turned into what many saw as a trumped-up vendetta against Larson himself. Several prominent figures in the academic and museum worlds who knew Larson took his side, citing the great care taken in the fossil lab at the Black Hills Institute, and the meticulous notes that were kept there. They called him a scientist. Others lumped him in with fossil hunters whom they regarded as little better than thieves, people who did a sloppy job of excavating and then sold their fossils to collectors, often foreign, thus preventing proper study by professionals. Robert Bakker, one of the country's foremost paleontologists (a leader among those who believe that dinosaurs were warm-blooded, like birds, rather than cold-blooded, like lizards), told Steve Fiffer that he had originally been suspicious of Larson but had changed his mind once he saw the lab at the institute. "We stiff Ph.D.s," he said, "speaking in Latin, wearing elbow pads, criticize the Larsons for not having degrees, but their research is better than ours."

There has always been a degree of tension between the academics and the field workers in paleontology. In the mid-nineteenth century, when interest in extinct animals first took hold, the vast majority of those who dug up the bones were amateurs. It was professionals like Richard Owen who became famous, however, because they did the subsequent scientific work that led to the classification and naming of extinct species. The great rivalry in America between Edward Drinker Cope and O. C. Marsh was further fueled by the fact that Cope did a lot of field work, while Marsh did very little. Both, in fact, hired many amateurs to search for fossils. Barnum Brown had considerable education, although not an advanced degree. It was field work that he loved, and he had a nose for fossils, a sixth sense like Sue Hendrickson's, that made him the most successful fossil hunter of his time.

Being a dinosaur hunter is akin to being a comet hunter, like David Levy (chapter 2). Levy is simply better at spotting comets than most professional astronomers, and the professionals have their hands full with work that absolutely requires an advanced degree. In terms of dinosaur hunting, the logistics and expense of mounting a major expedition, like the one Michael Novacek led for the American Museum of Natural History to the Gobi Desert, inhibit the number of such efforts. They take money away from other museum priorities. The so-called amateurs like Peter Larson operate on a shoestring. He lived in a trailer next to the institute, and it was little wonder that there were problems with the collecting truck on the day Hendrickson discovered "Sue": the vehicle was fifteen years old. Larson certainly wasn't getting rich from his work.

Most professional paleontologists wouldn't regard someone like Sue Hendrickson as a scientist at all, and she doesn't see herself as one, either—not exactly. She does know that she has a knack for finding things, important things that advance scientific knowledge in many cases, an instinct that is rare across the board, whether you are talking about amateurs or professionals. She doesn't think a degree would make her better at it. Indeed, she can give the impression that she suspects it might be inhibiting. She doesn't have to come up with results to justify the expenses of a museum-sponsored expedition. What's more, such expeditions go by the book—you don't go wandering off with your dog in the Badlands just because you have a hunch. As to the matter of knowledge, Hendrickson is capable of absorbing enormous amounts of new information very quickly. She was sure she had found a *T. rex* embedded in the sandstone cliff on Maurice Williams's ranch because she knew that the vertebrae of a *T. rex* were articulated in a particular way. That kind of knowledge can be acquired by anyone with a fine mind—it doesn't have to happen in a classroom. "Academics are snobs," she has said. Yet there are many academics like Robert Bakker who salute her. They know she has something extra, and that she doesn't care about credit, just the thrill of the find and the opportunity to add a little something to the accumulated knowledge of humankind about the vast mysteries of the natural world.

Sue Hendrickson has been asked, "How do you do it?" all her adult life, about rare fish, amber, and certainly about "Sue." She knows what the press wants to hear, and she isn't comfortable with the idea of a sixth sense per se. Nor does she see herself as blessed with some curious form of luck. "I don't believe in fate," she has said. Between the lines, one senses that the answer she would like to give is, "I can find things because I'm Sue Hendrickson," but that would sound self-important, and she is not that kind of person.

It is widely agreed that if Hendrickson had not found "Sue" that day, the specimen might never have been found, or at least not for decades, by which time it would have been weathered to the point that it would be much less complete. The cliff was located in badlands, not grazing area. Even Maurice Williams hadn't been near the cliff in years. What's more, unless someone knew how to look, the fossil could have gone unnoticed. Nestled among undulating weathered rock, it was not easy to see. You had to know to look for the dribbled-down bones at the base of the cliff. In addition, other fossil hunters might have ignored the area for a long time. "The Larsons have already been there," they'd say.

Nevertheless, there were those in the academic world who had it in for amateur fossil hunters of all kinds, and they egged on the prosecution in the case that was built against Peter Larson and his associates. The residents of Hill City, where the Black Hills Institute was located, including its mayor, backed Larson to the hilt. There were charges that the prosecuting attorney was using the case to further his own political ambitions—a common enough suspicion in many cases of many kinds. Peter Larson, Neal Larson, Terry Wentz, and three other associates were indicted on the day before Thanksgiving, 1993. There were 148 alleged felonies and 6 misdemeanors, ranging from conspiracy to money laundering and interstate transportation of stolen goods. In a chapter titled "You Can Indict a Ham Sandwich," Fiffer notes, "If convicted of all crimes, Peter Larson faced up to 353 years in prison and $13.35 million in fines."

The trial lasted seven weeks during early 1995, and the jury debated for two weeks. In the end, Peter Larson was convicted of only two felony charges, both relating to a failure to declare to Customs money taken in or out of the country; his brother Neal was convicted of a misdemeanor involving fossil collecting. One of the other defendants, Bob Farrar, who also worked at the institute, was convicted of two felonies for supposedly undervaluing, on Customs forms, two fossils brought into the country from abroad. Only Peter Larson would go to jail, finally sentenced to two years and a $5,000 fine. The sentencing, due to legal maneuvering on all sides, and influenced by a Supreme Court decision, did not take place until nearly a year later. In the meantime, Larson had been allowed to travel to Japan to mount an extremely successful dinosaur expedition. Larson could have received probation and a fine, as his brother did. The harshness of the sentence, many commentators felt, was due to the fact that the defense lawyer, Patrick Duffy, had gone too far in arguing the case in the press. There was little question that approach had succeeded with the jury, but it had angered the prosecuting attorney and the judge. Perhaps the most trenchant comment on the whole affair was the offense listed on Larson's Bureau of Prisons form: "Failure to fill out forms."

Susan Hendrickson was dragged into the trial, too. She had been off on new adventures in other parts of the world, diving in Australia for sunken galleons, and off Alexandria looking for the ruins of Cleopatra's palace. Back in 1992, she had testified before the grand jury that eventually charged Larson and his partners. She had been given a grant of immunity—her lawyer insisted that she take it because he thought the entire case was an irrational one. She finally agreed because she knew nothing that could harm Larson or his codefendants. Indeed, Hendrickson's testimony at the trial itself was beside the point and basically useless to the prosecution. Even so, she was sickened to be testifying for the prosecution, and deeply frustrated because she had been forbidden by the judge to communicate with Larson in any way. Larson understood her situation, and they have remained close friends.

Tyrannosaurus Sue's fate took even longer to decide. The initial suit claiming that she belonged to the Sioux had been complicated by government claims that under the terms of the Indian Reorganization Act of 1934 and the Antiquities Act of

1906, which were in many ways at odds with each other, the *T. rex* belonged to the federal government, not the tribe and not Maurice Williams. The courts finally awarded "Sue" to Maurice Williams, voiding his contract with the Black Hills Institute and leaving him free to get as much money for her as he could. He told the press he hoped to get a million dollars.

Williams decided to auction "Sue." This was not going to be any Midwestern country auction, however. Sotheby's of New York, the great auction house better known for dealing with Rembrandts and Picassos, was given the opportunity to operate in a whole new sphere, handling the sale not of a portrait of a seventeenth- or twentieth-century lady, but an entire lady of a certain age—the Cretaceous. On October 4, 1997, bidding began on Lot 7045. It was to start at a floor of $500,000. Larson and his Black Hills Institute had contributed to the auction catalogue and advised Sotheby's on the handling of the 65-million-year-old skeleton. A benefactor was going to be bidding on "Sue," hoping to bring her back to South Dakota and Larson's institute. South Dakota's governor, William Janklow, had declared, "She belongs in South Dakota. She lived here and died here, and we want her back."

Sue Hendrickson was at Sotheby's that day, fervently hoping that Larson would get her namesake back. Maurice Williams was there, hoping the *T. rex* from his cliff would bring at least a million. Peter Larson was not there. He had been released from prison after eighteen months but was still under a confinement rule that kept him at the Black Hills Institute in Hill City, South Dakota. The bidding lasted only a few minutes. When the gavel came down, "Sue" had been sold to the Field Museum of Natural History in Chicago for an incredible $8.36 million, including the Sotheby's commission. She went on display at the Field Museum on May 17, 2000.

Despite his prison sentence, Larson was as respected as ever in the world of dinosaur hunters, among the people of South

Dakota, and even, to his surprise, in the nation's capital. In 1998 he installed another dinosaur associated with the Black Hills Institute, named "Stan" after Stan Sacrison, with whom Larson shared credit for its discovery, at the Smithsonian Institution. Another ten *T. rex* specimens, some of them partial, none as resplendent as "Sue," have been found since 1990, and Larson's advice is often sought on dealing with them.

Susan Hendrickson has survived cervical cancer, first diagnosed in the same year she found her famous *T. rex*. She has a vascular condition in one leg that hampers her walking, although she can still dive with the best of them. Never a fan of the limelight, and having had all too much of it in connection with her great discovery, she has built a house in Honduras, where she can get away from the world. She goes on diving, looking for ancient treasures, wherever the opportunity arises around the world. What would she most like to find? Perhaps knowing what the press wants to hear, she says she would like to find another *T. rex*, but this time with a whole family of intact skeletons, not just the bits and pieces of juveniles discovered with "Sue."

Hendrickson once said that most people seemed to assume she must have a Ph.D. That amused her; she was in a way proud of being a high school dropout, someone who had learned on her own. But she couldn't escape academia entirely. In 2000 she was awarded an honorary doctorate from the University of Illinois at Chicago, for her "expansion of knowledge and performance of exemplary service." Another honorary degree followed from Barnard College in 2001. Academia has its own way of dealing with amateurs who are too remarkable to ignore.

To Investigate Further

Fiffer, Steve. *Tyrannosaurus Sue*. New York: W. H. Freeman, 2000. Fiffer's account of this story is an unusual example of fine journalism coupled with solid scientific background material. He manages to convey the excitement and hard work of dinosaur hunting, and to thread his way clearly through the subsequent thickets of legal complications.

Novacek, Michael. *Dinosaurs of the Flaming Cliffs*. New York: Anchor/Double-day, 1996. For those who prefer their dinosaurs straight, this book about a major expedition to the Gobi Desert is fascinating reading. It is both an adventure story and a serious consideration of the evolutionary issues pale-ontologists hope to settle by studying dinosaurs.

Norell, Mark A., Eugene S. Gaffney, and Lowell Dingus. *Discovering Dinosaurs*. New York, Knopf, 1995. The three authors, all of the American Museum of Natural History in New York, have put together one of the most complete books about the discovery, mounting, and significance of dinosaurs. Each of the fifty sections asks and answers a specific question about dinosaurs, mak-ing the book an easy-to-use reference. Dozens of photographs of dinosaurs in the museum's famous collection, as well as drawings and photographs from the field, illustrate this splendid book.

Psihoyos, Louie, with John Knoebber. *Hunting Dinosaurs*. New York: Random House, 1994. Praised by *Publishers Weekly* as "the ultimate book for dinosaur buffs," with lavish photographs by the photojournalist author, as well as drawings, maps, and historical photos, this book is packed with information and enormous fun.

Hendrickson, Susan, with Kimberly Weinberger. *A Dinosaur Named Sue: Hunt for the Past: My Life as an Explorer*. New York: Scholastic, 2001. Aimed at younger readers, this short autobiography celebrates the author's passion for knowl-edge and adventure.

CHAPTER 10

Felix d'Herelle

Bacteriophages Discoverer

It began with a plague of locusts in the state of Yucatan, Mexico. Felix d'Herelle, a self-taught French-Canadian bacteriologist, was working for the Mexican government in 1909 when locusts began attacking the crops in the area. The local people told d'Herelle that there were places where the "ground was strewn with the corpses of these insects," as he wrote in an article translated by J. L. Crammer and published in *Science News* in 1949.

> I went there and collected sick locusts, easily picked out since their principle symptom was an abundant blackish diarrhoea. This malady had not yet been described, so I studied it. It was a septicaemia [septic shock caused by a rapid fall in blood pressure] with intestinal symptoms. It was caused by bacteria, the locust coccobacilli, which were present almost in the pure state in a diarrhoeal liquid. I could start epidemics in columns of healthy insects by dusting cultures of the coccobacillus in front of the advancing columns: the insects infected themselves as they devoured the soiled plants.

Conducting a series of tests on the bacteria, d'Herelle made an observation that would later prove to be of great importance: "In the course of these researches I noted an anomaly, shown by some cultures of the coccobacillus which intrigued me greatly, although in fact the observation was ordinary enough, so banal indeed that many bacteriologists had certainly made it before on a variety of cultures. The anomaly consisted of *clear spots*, quite circular, two or three millimeters in diameter, speckling the cultures grown on agar." Noticing something and following up on it are two very different matters, however. Nineteen years later, Alexander Fleming would have a similar experience. Fleming noticed a mold growing in an uncovered petri dish at his laboratory; around the mold was a clear area. Analyzing the mold, he discovered penicillin, although it would be another decade before it became technically possible to produce the mold in sufficient quantities to test it as an antibiotic. In a paper on his discovery, however, Fleming suggested that it might have antibiotic properties, an aside that brought him the Nobel Prize along with the two men whose work made commercial production possible, Lord Florey and Ernst Chain. Like d'Herelle, Fleming suggested that others must have noticed the same kind of culture, but simply thrown it out as contaminated.

In 1909, d'Herelle did not carry out any further experiments on the anomaly he had noted, but he would return to it later when he had more time for pure research. Throughout his life, d'Herelle was a wanderer, taking jobs where he could get them and seldom staying put in one country or at a single institution for more than a few years. Born in 1873 in Montreal, Quebec, he was the only child of a wealthy French-Canadian father and a Dutch mother. His father died when he was six, and his mother then moved to Paris. William C. Summers of Yale University established in his 1999 book, *Felix d'Herelle and the Origins of Molecular Biology*, that d'Herelle's formal education never went beyond high school, but he had a brilliant mind, and his mother's largesse made it possible for him to travel widely. Little is known about his early life, and indeed there are holes even in the chronology painstakingly established by Summers. As a young man, d'Herelle developed an interest in bacteriology, and landed a job in Guatemala City as a hospital bacteriologist in 1901. Questions remain as to whether he passed himself off as a doctor or was simply assumed to be one on the basis of his knowledge. He did not treat patients, but taught at a medical school in Guatemala, and would later state that he gained much of his scientific grounding during the several years he spent in that country.

In 1907, d'Herelle moved on to Mexico, working on the fermentation of sisal, as well as serving as a government bacteriologist during a period when epidemics of yellow fever and other infectious diseases were widespread. Shortly after his encounter with the locusts, d'Herelle returned to Paris to work at the Pasteur Institute. He made several journeys abroad over the next few years, to Argentina, Turkey, and Tunisia, trying to use the methods of locust control he had employed in Mexico, dusting plants with the coccobacillus he had discovered. While in Tunisia, he again encountered the clear spots in cultures that he had first noticed in Mexico. Charles Nicolle, the director of the Pasteur Institute branch in Tunisia, later claimed to have suggested to d'Herelle that they were a virus.

Returning to Paris, d'Herelle was put to work at the Pasteur Institute on finding ways to control the deadly outbreaks of dysentery that were felling so many World War I allied soldiers. The cause of the dysentery was soon uncovered: *Shigella* bacteria. Once again, clear spots showed up in cultures of dysentery *bacillus* (rod-like bacteria). This time he followed through and investigated them thoroughly. He recognized that the clear spots, called *plaques*, were appearing because something was killing the bacteria. A series of tests using the feces of a particular patient produced more plaques on agar plates. (*Agar* is a jellylike carbohydrate from seaweed that is not broken down by microorganisms, and thus remains a solid neutral surface during the course of an experiment.) The next step was to take a plaque and mix it with a fresh culture of the bacteria in a flask.

"The next morning," he would write toward the end of his life, in the account published in 1949 in *Science News*, "on opening the incubator, I experienced one of those moments of intense emotion which reward the research worker for all his pains: at first glance I saw that the culture which the night before had been very turbid, was perfectly clear: all the bacteria had vanished, they had dissolved away like sugar in water. As for the agar spread, it was devoid of all growth and what caused my emotion was that in a flash I understood: what caused my clear spots was in fact an invisible microbe, a filterable virus, but a virus which is parasitic on bacteria."

D'Herelle gave his discovery the name *bacteriophages—phage* means "to eat" in Greek, thus giving us "bacteria-eaters." They are now generally referred to simply as phages. In his first paper on the subject, published in French in 1917, he gave a more sober account of the significance of his discovery. "In summary," he wrote at its conclusion, "in certain convalescents from dysentery, I have shown that the disappearance of the dysentery bacillus coincides with the appearance of an invisible microbe endowed with antagonistic properties with respect to the pathogen bacillus. This microbe, the true microbe of immunity, is an obligatory bacteriophage; its parasiticism is strictly specific, but if it is limited to the species at a given moment, it may develop antagonism in turn against diverse germs by accustomization." Here, d'Herelle pointed toward the work that would continue for the rest of his life: trying to find phages that could be used against a variety of bacterial diseases. During 1917–1918, d'Herelle noted thirty-four cases in which patients were cured of dysentery when the appropriate bacteriophages he supplied attacked the bacteria. If such applications of phages could be extended to other diseases, it could provide medicine with a potent new weapon.

Following the publication of this paper, it turned out that the English biochemist Frederick Twort had made the same discovery in 1915. The superintendent of London's Brown Institution, Twort began experiments with viruses to determine if they could thrive without a cellular host. He too used an agar plate. Although the virus didn't replicate, bacterial contaminants appeared over time and then became glassy in appearance and died. Twort published a brief paper on the subject of "glassy transformation," but it was little noticed. He might have continued his experiments, but was called up to serve in the military, and went into other lines of research after the war. It was d'Herelle's name for the virus that came into widespread use. The two men are generally given joint credit for the discovery, but d'Herelle's name takes precedence because his independent discovery was more fully described, and because he would go on to do much more work on phages, while Twort did not.

Independent discoveries of the same phenomenon by two different scientists have occurred again and again throughout the history of science. Isaac Newton and Gottfried Leibnitz invented calculus more or less simultaneously in the seventeenth century; Nikola Tesla gave a lecture on radio telegraphy in 1893, but Gugliemo Marconi, apparently ignorant of Tesla's work, came up with the same idea two years later, and since it was he who followed through and transmitted the first transatlantic signal in December 1901, he has received the credit for that invention. The Twort/d'Herelle case, however, became the basis of a famous novel. Sinclair Lewis's *Arrowsmith* concerns an American bacteriologist who discovers phages, only to find out that he has been beaten to publication by a researcher at the Pasteur Institute named Felix d'Herelle. The novel won the Pulitzer Prize for 1926, but Lewis refused it on the grounds that the prize should have gone to his earlier *Main Street* (1920) or *Babbitt* (1925). Lewis subsequently became the first American to win the Nobel Prize for Literature, in 1930.

While d'Herelle became famous enough from his discovery for Sinclair Lewis to latch onto his story, he was never entirely accepted by the scientific establishment. In part this was due to the fact that his credentials were always somewhat suspect (quite rightly, in academic terms), but also because he was a difficult man to get along with. His relationship with the Pasteur Institute was peculiar. Despite the publication of a full-length book on the use of phages in 1921, *The Bacteriophage: Its Role in Immunity*, he was never made a tenured member of the staff, and there was a serious rupture with the Institute around that time. Some accounts suggest that he was virtually booted out, perhaps because he was making what were regarded as excessive demands for phage research support. On the other hand, he seems to have maintained at least an informal relationship with the Pasteur Institute, since other scientists encountered him there as late as 1926.

In 1922, he began teaching at the University of Leiden, in the Netherlands, and was awarded an honorary medical degree by Leiden in 1924. But he was soon setting off for various tropical countries, usually at the invitation of governments, to see if he could develop phage-based antidotes to various disease outbreaks. He worked in Egypt in 1926, where a new onslaught of bubonic plague had occurred, and attempted to cure cholera in India in 1927. In his books he claimed considerable success with such ventures, if on a limited scale. But once he had left the scene, further success seemed to become elusive. He complained that those who were designated to follow up on his procedures failed to do so properly, which was entirely possible. This was a new and tricky field, in many aspects still in an experimental stage.

In 1928 he was appointed a tenured professor at Yale Medical School, with a salary of $10,000—then quite a princely sum—and laboratory support. Conflicts with other professors and the administration led to his resignation in 1933. Summers's portrait of d'Herelle suggests a man who was "bewildered" by the social niceties of the scientific establishment. To put it more bluntly, he seems to have been lousy at academic politics—which can be seen as a plus or a minus, in terms of character, depending on one's point of view. At any rate, he never fared well in establishment situations.

The next stop for the peripatetic d'Herelle was the Soviet Union. George Eliava, the director of a bacteriology laboratory in Tbilisi, Soviet Georgia, not far from Turkey, had encountered d'Herelle in Paris, at the Pasteur Institute, in 1926, and became fascinated with his work on phages. The two men hit it off personally as well, an unusual occurrence in d'Herelle's life. After leaving Yale, d'Herelle went to join the Eliava Institute in Tbilisi. Like many Western intellectuals of the time, d'Herelle had great hopes for the future of communism. In the preface of a book he wrote while at the institute, he said, "Only in the Soviet Union is it possible to condemn erroneous methods of prophylaxis, sanctified by time, without fear of insulting the habitual delusions having prescriptive rights. A new social system has the right to afford itself revision of everything that has been obtained as a heritage of the past." Clearly this was a broadside at Western medical professionals with whom d'Herelle had encountered difficulties.

This 1935 book, *The Bacteriophage and the Phenomenon of Recovery*, translated into Russian by Eliava, was dedicated to Stalin. That has led to controversy. There are those who believe d'Herelle had become an enthusiastic Communist, while other experts insist that such dedications had become almost pro forma, an aid to getting a book published. In fact, d'Herelle soon became alarmed at what was going on in the Soviet Union, as Stalin's lethal purges reached their height, and he left in 1936. Eliava himself was arrested in 1937 and executed. His institute's phage research continued, however, although the institute itself was merged with other laboratories and renamed several times. With the collapse of the Soviet Union, the laboratory became part of the Georgian Academy of Science. We'll come back to its continuing research later. Having returned to Paris, d'Herelle opened a commercial laboratory, which did well, but he was interned during the German occupation. His laboratory was reopened after the war, but he had only a few years left, dying in 1949. His discovery of phages would prove a crucial development in bacteriology, leading to a host of further discoveries and playing a large part in unraveling the functions of DNA. As we shall see, even d'Herelle's dream of using phages to combat disease has had a renaissance in recent years.

———

When d'Herelle originally discovered phages, he had referred to them as "invisible microbes." The electron microscope, invented in Germany in 1932, would change that, although it would be ten years before the first micrograph of a phage was made by Tom Anderson. The most common form of phages are those known as the T4 type. They make up about 95 percent of all phages. Under an electron microscope, they look something like a Lunar Lander, with an elongated pod or head, consisting of DNA encased in a protein wrapping, that sits atop six legs, or tails. When phages invade bacteria, they plunge their legs into the bacteria, transferring their own DNA into the bacteria. Eventually the bacteria will burst, releasing large numbers of additional phages, which will then feed on other bacteria. That is why phages can be used to combat disease. However, phages can also create more virulent, new strains of bacteria because of their ability to transfer genes from one microbe to another.

As phage research continued from the 1930s on, it was soon realized that there must be enormous numbers of them, particularly in seawater and freshwater environments. Not until the 1990s was an attempt made to count the number of phages by using a centrifuge and advanced electron microscope technology. In a milliliter of sea water, it was found that there were up to ten million tailed phages, and their presence in fresh water can reach a billion per milliliter. That suggests that, globally, phages may amount to 10 to the 30th power (a 1 followed by 30 zeroes).

Up until 1940, the main focus of phages research was to seek out those that would cure diseases, in a continuation of d'Herelle's original intent. Such research took place chiefly in the Soviet Union and in Europe. It was successful enough that the pharmaceutical giant E. Lilly began selling some phage's serums. As we noted earlier, however, a way was found to produce penicillin on a useful scale in the late 1930s. A crash program, with much of the work being done at the U.S. Department of Agriculture laboratory in Peoria, Illinois, produced sufficient penicillin to save innumerable lives during World War II. Large numbers of soldiers had died during World War I from infections rather than the wounds they had received. The development of penicillin and sulfa drugs essentially brought a halt to phage research on the disease control front, except in the Soviet Union, for decades to come.

By that time, however, an important new reason had been found for the study of phages. The prime mover in this new work was Max Delbruck, a scientist of truly protean abilities. He was initially interested in astronomy but then switched to quantum physics. His Ph.D. thesis was overseen by Max Born, who would win the 1954 Nobel Prize in physics for his work on the wave function in quantum mechanics. In 1931, Delbruck went to Copenhagen to become part of the famous group that surrounded Niels Bohr. Bohr, who had won the 1922 Nobel Prize for discovering the structure of the atom, was one of the most influential scientific figures of the century. He suspected that quantum physics might also have implications for biology, and that led Delbruck to change fields once again. Fleeing Hitler's Germany in 1937, he ended up at the California Institute of Technology, where the biologist Emory Ellis was working on phages.

Ellis had encountered Felix d'Herelle during the latter's five-year stint at Yale.

Delbruck then took a position at Vanderbilt University in Nashville, Tennessee. While there he met the Italian bacteriologist Salvatore Luria, who had also fled fascism to America and was at that point a research assistant at Columbia University. Delbruck was planning to spend the summer at the research and educational facilities at Cold Spring Harbor, New York. That was the beginning of what came to be called the "phage group," which expanded over the years to include many noted scientists with an interest in phage research, including several physicists who, like Delbruck, saw correspondences between genetic material and atomic structure. Even Leo Szilard, who had created the first nuclear chain reaction, became involved.

That first summer at Cold Spring Harbor, Delbruck and Luria established the "mutual exclusion principle" with respect to phages. It takes a set amount of time for each type of phage to cause bacteria to burst, or lyse, after attachment, almost to the minute. Delbruck and Luria compared two specific types, and found that one had a much faster lyse time compared to the other. But when the two types were plated to bacteria together, the slower type always succeeded in producing new phages. Two different types could not infect a bacteria simultaneously, and the type that was better suited to the job would win, excluding any action by the other, even though the winner took longer to lyse. Interestingly, d'Herelle himself had insisted that this was what happened, but he was not taken seriously on this point by establishment scientists, and his finding was not accepted until Delbruck and Luria proved it with more advanced techniques in 1941—even as d'Herelle himself was kept under internment by the puppet Vichy government of France installed by Hitler. If d'Herelle had known what was going on at Cold Spring Harbor, he might well have been indignant.

The following year, Delbruck and Luria returned to Cold Spring Harbor, as they would for the rest of the decade, and showed that phages were resistant to sulfa drugs, a fact that would have implications for the treatment of E. coli contamination and AIDS in later years. In 1943, Luria made a trip to Vanderbilt University to join Delbruck in some complex new experiments. This "fluctuation test" demonstrated that bacteria were capable of mutating in order to resist an attack by phages.

That same year, 1943, another major figure was added to the phage group at Cold Spring Harbor, Alfred Hershey of Washington University. Subsequent work on the genetic structure of viruses during the 1940s and into the early 1950s eventually brought the three scientists the Nobel Prize for Physiology or Medicine in 1969. The importance of the phage group lies not only in their specific work on the interplay between bacteria and phages, including both replication and mutation, but in the fact that they fully established the efficacy of phages as a basis for genetic research. The heads of phages, with their concentration of DNA surrounded by a protein wrapping, made them ideal for genetic experiments. It's important to understand, however, that at this point DNA was *not* regarded as the key to heredity. Indeed, we need to backtrack here, returning briefly to the nineteenth century.

In 1868, a year after Gregor Mendel published his paper on heredity in garden peas, as discussed in chapter 1, a Swiss biologist named Friedrich Meischer carried out a series of chemical tests on the nuclei of pus cells that he had retrieved from surgical bandages. He was able to detect an acidic substance that contained phosphorus. He called it *nuclein*. In addition, he detected a protein element within the cells that contained sulfur. Subsequently he found similar material in the heads of salmon sperm. The connection between his findings and Mendel's genetic studies of peas would not be recognized for decades, however.

By the end of the 1940s, the phage group's experiments, and those of other researchers, had made it clear that either the DNA or its protein wrapping must carry the genetic information that allowed replication. But which one was it? Most scientists were betting on the protein wrapping. The DNA portion was made up of only four subunits, whereas the protein portion had twenty subunits. Given the amount of information that needed to be passed on, it made more sense for it to be carried by the protein. To ascertain the truth of this assumption, in 1952 Alfred Hershey, now working with Martha Chase, used radioactive tracer isotopes—a new technology—in a famous pair of side-by-side experiments to determine which of the two carried the genetic message. In one experiment, the protein capsule was "labeled" with radioactive sulfur. In the second the DNA core was labeled with radioactive sulfur. (In 1868, it should be recalled, Meischer found that the nuclein contained phosphorus and the protein sulfur.) The two sets of radioactive phages were allowed to infect bacteria. Once that had taken place, two blenders were used to shake loose phage particles from the bacteria. These were tested and it became evident that the DNA fraction, with its radioactive phosphorus tag, carried the genetic code necessary to replication.

The climactic step was taken a year later. James Watson, an American geneticist whose Ph.D. thesis had been on phages, and who had worked at Cold Spring Harbor with the phages group, had gone to the Cavendish Laboratory at Cambridge University in 1950 for postdoctoral work. There he encountered Francis Crick, a brilliant physicist with a decidedly unorthodox mind. (Crick in later years would suggest that life did not arise spontaneously on Earth, but was a result of *directed panspermia*—in other words, that the planet was sowed with life by an advanced alien civilization.) Watson was in some ways an odd type himself, the sort of scientist who forgets to tie his shoelaces. They joined forces, determined to solve the mysteries of DNA, with the prospect of a Nobel Prize in their minds from the outset. Once again, we have a cross-fertilization between the fields of biology and physics at work. To fully understand how genes worked, it was necessary to understand the structure of genetic material, and physicists, having determined the structure of the atom, were old hands at this kind of conceptualization.

Watson and Crick were in no way starting at the bottom of the ladder. Much of the work necessary to their endeavor had already been done by others, but no one had been able to pull it together. In fact, Oswald T. Avery, Colin MacLeod, and Maclyn McCarty had published a paper in 1944 that suggested DNA was the bearer of genetic information. They had taken a strain of disease-causing, or virulent, *Streptococcus pneumonae* and shown that it could transform a nonvirulent strain into a virulent one. That seemed to prove that DNA carried the genetic material. But Avery was cautious about stating this conclusion too strongly, and because most scientists remained convinced that the more complex protein that encapsulated DNA must be the carrier of genetic information, the paper was to some extent shunted aside. However, the biochemist Erwin Chargaff of Columbia University was intrigued and, following the lead of the Avery group, showed in 1950 that even though DNA was identical from one cell to another in a given species, it was quite different from species to species. This suggested that DNA was more complex than had been believed.

Watson and Crick used the work done by Chargaff, by the phage group, and by Hershey and Chase, put it together with Linus Pauling's work on the structure of proteins (for which he would receive the 1954 Nobel Prize for Chemistry), and made important use of the X-ray diffraction studies of Rosalind Franklin and Maurice Wilkins at King's College, London. They had all the pieces of a large puzzle laid out before them. All they had to do was put them together—except that the pieces seemed to change size and shape whenever they were looked at from a new angle. In fact, the first model of the structure of DNA they put together was wrong. As Watson would later recount in his famous book *The Double Helix*, he was explaining what he had come up with, and how beautiful it was, to the American physical chemist and crystallographer Jerry Donohue, who happened, by the luck of the draw, to be sharing the same office as Watson and Crick. Donohue said they'd got it wrong, that the published tautomeric formulas were unsupported by good evidence, and he suggested a different approach. Watson had made cardboard models to work with, but when he changed to Donohue's specifications, the results didn't mesh at all. He began playing with them again the next morning, the way one turns the pieces of a puzzle around on a table, and suddenly he saw it: the elegant double helix structure of DNA (two strands that run in opposite directions and are complementary) lay on the table before him.

With the publication in 1953 of the DNA structure, the fundamental mystery was solved. DNA, and not protein, carried the genetic messages necessary for replication. It was now possible to begin the research that, nearly a half-century later, has given us a basic map of the human genome. Watson, Crick, and Wilkins would receive the 1962 Nobel Prize in Physiology or Medicine. Rosalind Franklin would rightfully have been included, but she had died, still in her forties, of cancer, and posthumous Nobel Prizes are not given. If they were, it might well have been argued that Felix d'Herelle should have received one. His work on phages underlay the entire development of molecular biology. But that is not the end of the story, either. His primary interest was in the use of phages to combat disease, and now, more than a half-century after his death, important new developments are occurring in that very field.

———

The development of penicillin and sulfa drugs saved millions of lives over the ensuing decades. But by the 1970s, patients were demanding that their physicians prescribe antibiotics for the common cold. Colds are caused by viruses, and antibiotics have no effect on them. Such overuse of antibiotics has had the disastrous result of giving some lethal bacteria the opportunity to mutate in ways that make them as impervious to most antibiotics as viruses are. There have been papers in medical journals and occasional articles in the mainstream press concerning this problem for nearly a decade. But only recently has the problem hit the headlines—the clinic I go to has copies of a local paper with a recent banner headline "OVERUSE OF ANTIBIOTICS CAUSES CRISIS" posted in all its examining rooms. Unfortunately, we have indeed reached a crisis point. While the major drug companies are working on a new generation of antibiotics, none are expected to be approved for general use until 2003 at the very earliest. And even these new drugs may not be capable of fighting some of the most dangerous bacterial infections. The World Health Organization (WHO) reports that penicillin has been rendered useless against almost all strains of gonorrhea in Southeast Asia, while in India typhoid species have developed a resistance to the drugs that were previously regarded as the most effective against that disease. Nor is the problem limited to developing nations. WHO estimates that 14,000 deaths each year in the United States are attributable to antibiotic-resistant bacterial infections, many of them acquired during hospital stays for unrelated treatment. Tuberculosis cases have been multiplying rapidly in American cities because new strains are resistant to antibiotics. In an October 2000 article in *Smithsonian Magazine*, Julie Wakefield quotes Alexander Sulakvelidze of the University of Maryland as saying, "Modern medicine could be set back to its pre-antibiotic days. It's a biological arms race."

Suddenly, phage treatments for a host of diseases are once again being seriously considered. Sulakvelidze is an expert in the field who often worked with the Eliava Institute in Tbilsi. There, from d'Herelle's time on, phage cultures were gathered from the Volga River. In Maryland, they are collected from Baltimore's inner harbor. As we have seen, phages exist in astronomical numbers in both fresh and salt water. There is no lack of supply of raw material. But phages are finicky "eaters." They will only devour very specific bacteria, and making the proper match entails a great deal of hit-or-miss laboratory work. Because phage research continued without a break throughout the twentieth century in Russia and at some other European laboratories, despite the development of penicillin and the sulfa drugs, the expertise and experience of foreign bacteriologists has brought many of them to the United States in the past few years as American universities have begun research in the field.

New companies have also been founded, such as Intralyx, co-founded in Baltimore by Sulakvelidze, and Phages Theraputics near Seattle, Washington. The British newspaper, *The Guardian,* reported in 1999 that the latter company had had an early success in curing a Canadian woman who was on the verge of death from an antibiotic-resistant infection. This story quoted Martin Westwell of Oxford University, who noted, "Because the virus is a living thing, every time the bacteria take a step forward in evolution, a step forward in the arms race, the virus can take its own evolutionary step forward."

An ABC radio interview with a number of experts in April 2000 made clear how seriously phage therapy is now being taken by a number of American researchers. One of these is Betty Kutter of Evergreen State College in Washington. She had worked with phages since 1963, but only in terms of DNA research. In 1980, however, a grant took her to Tbilisi. A patient who had had a leg infection for months had reached the point where it had been decided it would be necessary to amputate. But phage therapy was tried first. "First they split his foot open in the area where the wound was, halfway back from between the toes, and they had put in phage a few days before I was there, and I was there in the operating room when they opened it up for the first time after that, and the wound was completely clean, it was really very amazing." When she returned to the United States, Kutter became head of a nonprofit group called

PhageBiotics, which has been doing important research on phage therapy.

Progress has been made, but developing phages that will work in human beings presents problems. Even d'Herelle had been frustrated by the fact that phages proved to be so specific to given bacteria. One that worked beautifully in respect to a given type of bacteria would have no effect against others. In addition, the specificity of phages makes it difficult to be certain that the tests on animals that are required by the FDA can be considered convincing evidence that they will be safe when used on human beings. Nevertheless, the FDA has looked kindly on phages research because it recognizes that a great many people are dying of antibiotic-resistant bacteria.

Richard Carlton of Exponential Biotherapies explained one major cause of death to ABC's Richard Aedy. *Enterococcus faecium,* a common bacteria found in the human stomach, causes no problems for most people, but people with immune deficiencies can be killed by it. In such cases, the bacteria has proved resistant even to vancomycin, the antibiotic of last resort. Called vancomycin-resistant enterococcus, it first appeared in 1990. Carlton explains that in hospitals, "about 60 percent of all the strains of *Enterococcus faecium* are now vancomycin resistant, which means untreatable, and this little beast, this little bacterium, is so hardy it survives on stethoscopes, EKG knobs, it's all over the hospitals and they can't kill it with antiseptics. So it just gets spread around and spread around, and virtually colonizing hospitals." Carlton's company has developed a phage that has been 95 percent effective against this new strain. It went into clinical trials in late 2000, and while those can last for years, the FDA is so concerned about the situation that it may allow the process to be sped up.

There are numerous other uses for phage therapy under development. Some could be on the market by 2003, just at the point when many experts expect the antibiotic-resistant bacteria crisis to reach a peak. Richard Aedy reported that George Poste, the head of research at the pharmaceutical giant GlaxoSmithKline, has suggested that we could reach a point where people could get a sore throat on Tuesday and be dead by Friday. In such a crisis, phage therapy could save an enormous number of lives. There are doubters, however, including Ian Molineux, a microbiologist at the University of Texas in Austin, who feel that while there may be some major successes with phage therapy, it's unlikely to achieve the widespread uses of antibiotics, simply because of the fact that a different type of phage is needed for each kind of bacteria. Others maintain that phage research is entering an entirely new stage made possible by advances in biotechnology.

Regardless of how great the eventual importance of phage therapy proves to be, an entirely new chapter—perhaps even a full-length sequel—to the story of Felix d'Herelle's discovery is now being written. An odd, somewhat irascible, and largely self-educated bacteriologist who took note in 1909 of mysterious clear spots on an agar plate while dealing with a plague of locusts opened up an extraordinary new line of biological research. The bacteriophages he named in 1917 proved crucial to the eventual discovery of the structure of DNA. That knowledge led to a biotechnology industry that holds out the promise of altering not only the way we live, but, through genetic manipulation, the way we are born. Although d'Herelle died four years before Watson and Crick's great breakthrough, he knew what the goal was. Even though the scientific establishment had in many ways given him short shrift, he could comfort himself with the fact that they were using his discovery as a key to a door that would open upon vistas almost unimaginable when he began his study of phages. That is certainly a legacy any scientist, amateur or professional, would be proud to claim. But he might be even more satisfied by the developments in phage therapy that have taken place in the past decade, which lie closer to his own original dream. Phages, he believed, could eradicate some of the most ancient of human diseases. He may yet be proved right.

To Investigate Further

Summers, William C. *Felix d'Herelle and the Origins of Molecular Biology*. New Haven: Yale University Press, 1999. This is the only full-scale biography of d'Herelle, researched over many years. It is a splendid example of the kind of book that reminds us how many fascinating individuals there are who remain little known to the general public.

Watson, James D. *The Double Helix*. New York: Athenuem, 1968. This celebrated book, a major best-seller when it was published, is one of the most accessible books ever written by a scientist intimately involved with a great scientific breakthrough.

Tagliaferro, Linda, and Mark V. Bloom. *The Complete Idiot's Guide to Decoding Your Genes*. New York: Alpha Books, 1999. As noted in the first chapter, this is an excellent introduction to the subject of genetics for the general reader.

Note: There is a great deal of material on bacteriophages to be found on the Internet. Yahoo.com, Google.com, and others will point you to many possible links, simply by entering the search term "bacteriophages." There is considerable duplication of information from site to site, but it is also more reliable than is often the case with more widely popular subjects.

INDEX